Predators

Also by Brian Glyn Williams

Afghanistan Declassified: A Guide to America's Longest War

*The Crimean Tatars: The Diaspora Experience
and the Forging of a Nation*

*The Last Warlord: The Life and Legend of Dostum, the Afghan Warrior
Who Led US Special Forces to Topple the Taliban Regime*

Related Titles from Potomac Books

A History of Air Warfare
—John Andreas Olsen, ed.

Imperial Designs: War, Humiliation & the Making of History
—Deepak Tripathi

*Killing without Heart: Limits on Robotic Warfare
in an Age of Persistent Conflict*
—M. Shane Riza

The Pakistan Cauldron: Conspiracy, Assassination & Instability
—James P. Farwell

Predators

The CIA's Drone War on al Qaeda

BRIAN GLYN WILLIAMS

Potomac Books
Washington, D.C.

© 2013 by Brian Glyn Williams
All rights reserved
Potomac Books is an imprint of the University of Nebraska Press

Maps courtesy of Cuango Liu.

Library of Congress Cataloging-in-Publication Data
Williams, Brian Glyn.
    Predators : the CIA's drone war on al Qaeda / Brian Glyn Williams. — First Edition.
        pages cm
    Includes bibliographical references and index.
    ISBN 978-1-61234-617-5 (hardcover : alk. paper)
    ISBN 978-1-61234-618-2 (electronic)
    1. Qaida (Organization) 2. Terrorism—Religious aspects. 3. Drone aircraft—United
States. 4. War—Moral and ethical aspects—United States. I. Title.
    HV6431.W5655 2013
    327.127305491'1—dc23

                                                                                    2013006662

Printed in the United States of America on acid-free paper that meets the American
National Standards Institute Z39-48 Standard.

Potomac Books
22841 Quicksilver Drive
Dulles, Virginia 20166

First Edition

9 8 7 6 5 4 3 2 1

For my wife, Feyza

CONTENTS

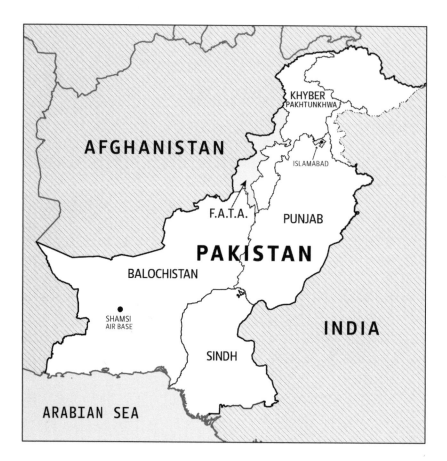

FEDERALLY ADMINISTERED TRIBAL AGENCIES IN PAKISTAN

PREFACE

At the dawn of the twenty-first century, the United States, the world's lone remaining superpower, launched itself across the globe in a war against terrorists who had struck at its heart on September 11, 2001. Among the most novel aspects of this campaign was the use by an ostensibly civilian organization—the Central Intelligence Agency (CIA)—of a new generation of unmanned, remote-control planes known as drones to kill the enemy hiding out in remote sanctuaries in Pakistan, Yemen, and Somalia. (The military has been in control of most drone operations in Iraq and Afghanistan.) Thousands of people, both terrorists and civilians, have been killed and many more continue to die in this aerial assassination campaign, yet there has been no thorough study of this remote-control warfare in remote lands to date. Critics and supporters of the drone killings debate the issue, often passionately, without having a basic outline of the vast covert operation.

As a historian I have made an effort in this book to fill this void, to record the history of what amounts to an all-out CIA drone war on the Taliban and al Qaeda in order to inform the debate on this controversial topic. In essence this book takes hundreds of disparate reports of drone attacks and weaves them together into one narrative. This work is not a polemic designed to cast moral aspersions on the campaign nor is it meant to condone it. Neither is it a study about the ethics and morality of this new cutting-edge killing technology. It is purely a record of these extraordinary events for the generation experiencing history's first remote-controlled drone war. Proponents and opponents of the campaign can do with this story what they will.

But first a word of caution: Because the CIA, the primary organization that runs the drone operations in Pakistan, Yemen, and Somalia, rarely acknowledges this covert counterterrorism campaign, this book is based entirely on open-source documents from the West and Pakistan. These sources can be flawed, but when taken together they are incredibly useful in providing new insights and the first overview of this murky chapter in the war on terrorism.

ACKNOWLEDGMENTS

Many thanks to my parents, Gareth and Donna, who inspired me to study Muslim Eurasia in my youth and have continued to support me up until today. Thanks also to Feyza, my forbearing wife, for putting up with my obsession with drones these last few years. *Teshekur* (thanks) are also due to my parents-in-law, Feruzan and Kemal Altindag, who once again gave me a quiet place to write in their house on the coast of Turkey. I would also like to express my gratitude to Rafay Alem and his wife, Aysha, for letting me stay with them in their home in Pakistan. Tim Paicopolos did a masterful job of editing this volume, so I also owe him a debt of gratitude.

In this work I cite a drone study carried out in conjunction with my colleagues at the University of Massachusetts–Dartmouth, Avery Plaw and Matthew Fricker, so thanks to them as well. Thanks also to my chairmen/colleagues Len Travers and Mark Santow for providing me with a supportive, collegial environment in which to write this book. This work also benefited from the drone studies of Bill Roggio and Alexander Mayer at the indispensible *Long War Journal*, Peter Bergen and Katherine Tiedemann at the New America Foundation, and Noah Shachtman and other writers at *Wired*, so thanks to them for their excellent research. And *shukriya* to my Pashtun driver, Saki, who risked his life driving me through the tribal zones of Pakistan. Thanks to Canguo Liu for creating the wonderful maps found in this book and to my colleague at the University of Massachusetts–Dartmouth, Spencer Ladd, for recommending Liu. I would also like to thank Jason Seto for the constant barrage of stories related to drones he has sent me over the years. Lastly I would like to thank my former PhD advisers, Uli

Schamiloglu and Kemal Karpat, for guiding me in my journey to understand the history of Muslims in Afghanistan and surrounding regions.

1

The Death of a Terrorist

Soon we will launch an attack in Washington
that will amaze everyone in the world.
—**Pakistani Taliban leader Baitullah Mehsud**

It is perhaps the worst-kept secret in the war on terror. In 2004 the Central Intelligence Agency (CIA) launched what amounts to an all-out airborne war, its most extensive assassination campaign since the Vietnam War, against Taliban and al Qaeda members hiding out in Pakistan's wild tribal zones. Thousands have been assassinated in this covert bombing campaign, which is being waged by remote-control drones, or unmanned aerial vehicles (UAVs), known as Predators and the more advanced Reapers. These high-tech weapons in the sky have indisputably killed hundreds of al Qaeda and Taliban leaders who are actively planning new terrorist attacks on the American homeland or on Coalition troops in neighboring Afghanistan. Whether acting in a "force protection" role against Taliban insurgents trying to cross the Afghan-Pakistani border and attack U.S. and Coalition troops operating in Afghanistan or attempting to disrupt al Qaeda's future acts of mass-casualty terrorism, the UAV drones are merciless and efficient killers. The enemy never knows when or where the *machays* ("wasps," as they are known in the local Pashtun language) are going to strike next and lives in constant fear of being incinerated by Hellfire missiles fired by high-flying UAVs that are barely visible from the ground.

With their ability to loiter for up to twenty-four hours and use high-resolution cameras to follow their targets from afar, the UAV drones have added a level of precision to this bombing campaign that has never been seen in previous aerial campaigns. Whereas in earlier bombing campaigns high-flying jets dropped clumsy, unguided five-hundred-to-two-thousand-pound "dumb bombs" on their targets, the slower propeller-driven Predators and Reapers hover over their targets, track their "pattern-of-life movements" with high-resolution and infrared cameras, and fire smaller missiles and mini-bombs that are guided by lasers or satellites.

But as precise as they may be, the drones' Hellfire or Scorpion missiles and Paveway guided bombs have also killed civilian bystanders. Scores of Pakistani tribesmen whose only crime was to be near a targeted al Qaeda convoy or Taliban *hujra* (compound or guest house) have been killed as unintentional collateral damage. The killing of Pakistani citizens on Pakistani soil by distrusted foreigners has, not surprisingly, caused a backlash of anti-Americanism in this proud country. The paradox then is that America's most effective tool in killing high-value al Qaeda and Taliban targets may also be driving average Pakistanis to see the United States as their enemy. This could undermine the unstable pro-American government in this sprawling nuclear-armed country of 190 million people and inadvertently help recruit new terrorists.

SOUTH WAZIRISTAN AGENCY, PAKISTAN, AUGUST 5, 2009

This conundrum is best illustrated by the drone hunt to kill Pakistan's most wanted man, Baitullah Mehsud. Mehsud was the leader of a coalition of Pakistani Taliban groups known as the Tehrek e Taliban e Pakistan (TTP). In 2007 Mehsud declared war on Pakistan and began to send waves of suicide bombers against Pakistani targets from his remote hideout in the mountainous tribal zones of Pakistan's northwestern border with Afghanistan.[1] These suicide bombers eventually killed thousands of Pakistanis.[2] In one year alone (2009) Pakistani sources claimed that more than three thousand Pakistanis were killed by the Taliban.[3] The murderous campaign was a slow-motion version of al Qaeda's 2001 attack on the United States, and by 2008 Pakistan had surpassed Iraq as the number-one target for suicide bombers. No one seemed safe from the Pakistani Taliban terrorist mastermind who had no compunction about deliberately killing men, women, and children in his war against the "pro-American puppet" government of Pakistan.

Among Mehsud's most famous victims was the former president of Pakistan, the wildly popular Benazir Bhutto, who had just returned from exile bravely promising to stand up to the Taliban "cancer" that was devouring her country. Although many conspiracy-oriented Pakistanis have blamed everyone from the Israelis to the Pakistani government itself for killing Bhutto in December 2007, Mehsud had loudly promised he would kill her if she returned to Pakistan, and he appears to have fulfilled his promise.[4] Upon hearing of her death at the hands of his assassins, he is reported to have gloated, "Fantastic job. Very brave boys, the ones who killed her."[5]

But Bhutto was not the only one to die in Mehsud's bloody terror campaign. Mehsud's fedayeen (suicide bombers) entered holy Sufi-mystic shrines, hospitals, factories, anti-Taliban jirgas (tribal meetings), mosques, Pakistani army and Inter Services Intelligence (ISI) bases, a Marriott hotel, polling stations, police recruitment posts, political rallies, refugee camps, and other public places and detonated themselves, slaughtering and maiming countless civilians. The unprecedented slaughter was a shock to many Pakistanis who had previously been tolerant of the Taliban. Mehsud's objective seemed to be to shatter Pakistani civil society just as the insurgents had previously done in Iraq.

Although the Pakistani army made several halfhearted efforts to enter the untamed mountainous region from which Mehsud ran his terrorist state in the tribal province of South Waziristan, they were seemingly incapable of conquering his rugged realm. In fact, Mehsud's hardy Taliban fighters beat the Pakistani army in several battles, on one occasion capturing more than two hundred Pakistani soldiers and beheading several of them on video.[6] One of the oppressed people of South Waziristan claimed, "South Waziristan now seems like a state within the state, and Baitullah Mehsud is running this like a head of government. Now he's an all-powerful man whose writ and command is visible across the tribal belt."[7]

In his breakaway western realm Mehsud's followers began to enforce a strict version of Islamic shariah law that was similar to the draconian system enforced by the Taliban in neighboring Afghanistan prior to 2001's Operation Enduring Freedom. Movie theaters, DVD stores, and girls' schools were burned, television sets were destroyed as "satanic devices," women caught in adultery (often something as innocuous as being caught in public with a man who was not their husband) were publicly stoned to death, those accused of stealing were arbitrarily executed, tribal leaders known as *maliks* were killed, men were forced to grow long Taliban-style beards, and a gloom settled on the province of South Waziristan and neighboring tribal regions conquered by the Pakistani Taliban. In essence, the secular system of Pakistan's founder, Muhammad Ali Jinnah, and the laws of Pakistan had been overturned in Mehsud's mini-Talibanistan. The darkness of the Afghan Taliban had simply migrated across the border to the Pakistani tribal zone once the Americans invaded Afghanistan.

For many Pakistanis who saw South Waziristan and the other wild Pashtun tribal lands on the northwestern frontier as an autonomous realm (known as the Federally Administered Tribal Agencies, or FATA) that had not really been a part of Pakistan proper, the de facto secession of this territory was not alarming. India, not fellow Muslims, was after all perceived as the main threat to Pakistan, regardless of how fanatical those Muslims might be. The "Talibanization" of Pakistan alarmed the Americans, whose newly elected president, Barack Obama, described this process of radicalization as "a cancer that risks killing Pakistan from within" more than it harmed the somnolent Pakistanis who had previously sponsored the Taliban.[8]

But the Taliban militants were not satisfied with terrorizing average Pashtun tribesmen in the FATA or carving out fundamentalist theocracies in the remote tribal agencies, and they began to move from their autonomous border provinces into Pakistan proper in 2008. Their tribal followers rose up in the scenic mountain province of Swat Valley, which is located about a hundred miles to the northwest of the Pakistani capital of Islamabad. In this former tourist resort area, they similarly carved out a Taliban-style militant state and began beheading police, burning girls' schools and beating girls on video, and enforcing strict Islamic law.[9]

The videos of the unbearably gruesome beheadings and floggings went viral in Pakistan and finally began to disturb average Pakistanis who had previously been strangely willing to overlook the Taliban's brutality.[10] The violence had come too close to home. By 2009 the Taliban had begun to spread from the Swat Valley to neighboring Buner Province. No one knew where the terrorists would stop. For the India-obsessed Pakistanis, Mehsud and his Taliban vigilantes had finally come to be seen as a national threat.

At this time the desperate Pakistani government began to ask the United States to use its Predator and Reaper drones, which had been targeting al Qaeda and exiled Afghan Taliban leaders in Pakistan's tribal zones, to kill the man Pakistani officials described as "the mother of all evil."[11] The Americans were only too happy to oblige their allies and had reasons of their own to seek the demise of Pakistan's most wanted man. Whereas Mehsud's Pakistani Taliban were mainly engaged in terrorism and war against the Pakistani state, one of the group's subcommanders, Hakimullah Mehsud, had been attacking U.S. supply convoys traveling through the nearby tribal agency of Khyber to bring supplies to American and Coalition troops serving in Afghanistan.[12] Plus, the Americans had been stung by a persistent rumor among paranoid Pakistanis that the United States was somehow sponsoring Baitullah Mehsud to create an excuse for conquering their state and gaining access to their prized nuclear weapons.[13] If this were not enough, Mehsud had also promised, "Soon we will launch an attack in Washington that will amaze everyone in the world."[14] Washington and Islamabad agreed that something had to be done about Mehsud.

Thus began the hunt for the notorious Taliban mass murderer. American troops were not allowed on Pakistan's sovereign territory so the drones were the obvious choice for taking Mehsud out. According to one study, more than a dozen U.S. drone strikes were eventually conducted against Mehsud and his followers once it was decided to assassinate him. But the elusive Mehsud seemed to be impossible to kill.[15] He moved from *hujra* to *hujra* and rarely gave interviews to outsiders as some of his more media-savvy comrades were known to do. Finding Mehsud in the autonomous mountainous tribal agency of South Waziristan, which was inhabited by almost a half million Pashtun tribesmen, was a Herculean task.

Then, on June 23, 2009, the CIA caught a break when it learned that Mehsud would be attending a funeral in the village of Najmarai in the Makeen District of South Waziristan to commemorate the earlier drone assassination of one of his top lieutenants, named Niaz Wali. CIA drones were scrambled to the scene and sent images to their U.S.-based remote pilots from their high-resolution cameras of the crowd gathered at the funeral. Pakistani spies working for the CIA indicated that the notoriously reclusive Mehsud had blundered and would indeed be in attendance at the funeral along with some of his top commanders. He was somewhere in the crowd of militants and villagers who had gathered to provide a Muslim burial for Niaz Wali and several other slain Taliban fighters. Rarely had so many Taliban leaders gathered in one place at the same time. Although the CIA had previously been comparatively selective in its targeting for fear of killing bystanders and upsetting their Pakistani allies, the decision was made to launch an attack on the funeral. Mehsud was too high value a target for both the Pakistanis and Americans to let him escape. No one in the CIA knew when they would have the chance to kill him again, so the order to fire was given.

Within minutes the remote pilots' screens back at CIA Headquarters in Langley were filled with the images of explosions as the drones' AGM-114 Hellfire antiarmor missiles slammed into the gathered crowd. As the smoke cleared, the CIA drone operators would have doubtless seen many "squirters" (i.e., survivors fleeing the explosions) as well as numerous dead and dying people lying scattered around the detonation zone (known as "bugsplats" in CIA parlance). But the drones were not done. According to one source, the drones subsequently fired missiles at Taliban members attempting to flee the scene of the attack in sport utility vehicles (SUVs), killing several more.[16]

Having decimated the funeral, the remote-control drones then flew back to their bases, located to the south, in the Pakistani province of Baluchistan, and the CIA waited for human intelligence (humint) to tell them whether they had killed Mehsud or any of his top leadership.

It did not take long for reports to come from the agency. Even though Pakistani and foreign journalists were denied access to this remote region, word began to trickle out that between sixty and seventy people had been killed in the missile strikes. This was the second deadliest drone strike in the ongoing campaign (the first took place in 2006 in the town of Damadola).

Then came the unfortunate news: for all the mayhem, Mehsud was not among those killed. Minutes before the strike, he had left the funeral and thus avoided the fate of several of his top commanders, who were blown up by the missiles. Mehsud had reportedly been so close to the explosion that it had damaged his car.[17] Among those who were not so lucky was a notorious leader named Sangeen Khan, who was the Pakistani Taliban's top commander in Afghanistan.

Members of the Taliban were not the only ones killed. According to one Pakistani source half of those killed in the strike were villagers.[18] Another local source claimed that of the sixty-seven people killed in the explosion eighteen had been villagers.[19] For their part, the Taliban, which had cause to inflate the number of dead civilians, claimed that only five of its members had been killed; the rest of the casualties were bystanders.

As with most of the CIA's strikes on the Taliban, it was difficult to nail down the details of who had died in this remote zone. One eyewitness who lost a leg in the strike reported, "After the prayers ended people were asking each other to leave the area as drones were hovering. First two drones fired two missiles, it created havoc, there was smoke and dust everywhere. Injured people were crying and asking for help. . . . They fired the third missile after a minute, and I fell on the ground."[20]

Regardless of what portion of the victims comprised civilian bystanders, it was the second highest death toll from a drone strike to date. Surprisingly, although there had been protest marches against previous drone strikes—most notably after the aforementioned Damadola strike—on this occasion there seemed to be little outrage in Pakistan at the killing of so many people at a funeral. Many in the pro-Taliban Islamist political parties grumbled about the sacrilege of attacking a funeral, but by this time most Pakistanis were willing to countenance the unpopular CIA drone strikes if it would rid them of Mehsud.

So the drone hunt continued. In the succeeding weeks and months the Pakistanis worked hand in glove with the CIA to track down the elusive Mehsud. The CIA launched several strikes on Mehsud's followers in an effort to kill him, but he never seemed to be at the scene of the attack.

Then, in August 2009 the CIA got lucky. Word came out of South Waziristan that Mehsud had married a second wife with the aim of having a male child after his first wife had given him only daughters. CIA and ISI spies determined who the woman was and found that she was the daughter of local cleric named Ikramuddin Mehsud. The trackers now had the scent of the prey. On the night of August 5, the CIA learned that Baitullah Mehsud, who was a diabetic, had traveled to his father-in-law's house in the village of Zanghara because he was feeling ill.

When a drone was sent to the scene, the CIA pilots flying the plane from seven thousand miles away in the United States were shocked to see images of several people gathered on the roof of the father-in-law's house. The man in the center of the crowd was receiving a glucose drip while his bodyguards looked on. Mehsud was known to be a diabetic; the man had to be him. This was the opportunity of a lifetime for the CIA pilots. The order was given to fire Hellfire missiles. Once again the CIA screens in distant America were lit up with explosions as the precision-guided missiles slammed into the unsuspecting people on the clay house below. The ultimate combination of humint and technological intelligence (techint) had called forth a "decapitation strike."

Then the smoke cleared. Although the CIA pilots could not be sure, it looked as if everyone on the roof, including seven bodyguards, Mehsud's new wife, and Baitullah Mehsud himself, was dead. Having expended its ammunition, the Predator drone turned and flew back to its base in the south while its handlers awaited news from distant Pakistan on the fate of the target.

It did not take long for word to emerge from South Waziristan. On the following day rumors began to spread that someone important had indeed been killed in the strike—Mehsud's wife. By that evening newspapers around the globe were publishing stories about the killing of the terrorist mastermind's wife.[21] Panicked Pakistanis feared a wave of suicide bombings as revenge for her death while Taliban leaders denied that Mehsud himself had been killed.

Then word of a large funeral to be held in the village the next day trickled out from the Taliban-controlled territory. A woman would never merit such an honor among the conservative Pashtun tribesmen of South Waziristan. It had to be someone more important than Mehsud's wife. A Taliban spokesman delivered the stunning news: With great sadness he announced that Mehsud had achieved "martyrdom." Pakistan's most wanted man was finally dead. Almost simultaneously a Pakistani military spokesman announced that he had actually seen the kill video filmed by the very drone that had fired on Mehsud. The Pakistani source claimed, "This is one hundred per cent. We have no doubt about his death. He is dead and buried." According to this source, "He was clearly visible with his wife. And the missile hit the target as it was. His torso remained, while half of the body was blown up."[22]

There were no public outcries from Pakistanis about CIA violations of sovereignty on this occasion. On the contrary, many Pakistanis secretly celebrated. The Pakistani newspaper *Dawn* ran a headline celebrating the death of Mehsud that read, "Good Riddance, Killer Baitullah."[23] One blogger from the Pakistani port city of Karachi claimed, "If [his death is] true, it would be good news and shows the value of drone attacks," and another wrote, "The mass murderer has met his fate. He was responsible for the death of thousands of innocent Pakistanis. May he burn in hell for eternity."[24]

The Americans were no less jubilant. President Barack Obama, who had stepped up the drone attacks soon after taking office, announced with grim satisfaction that the United States had "taken out" the terrorist chief. White House Press Secretary Robert Gibbs said, "Baitullah Mehsud is somebody who has well earned his label as a murderous thug. If he is dead, without a doubt the people of Pakistan will be safer as a result."[25] Roger Cressey, a former counterterrorism official on the National Security Council, said, "Mehsud was someone both we and Pakistan were happy to see go up in smoke."[26]

But those who thought the Pakistani Taliban had been beheaded by Baitullah's death were to be disappointed, for the terrorist group quickly held a *shura* (council meeting) and chose as Baitullah Mehsud's successor the fearsome Hakimullah Mehsud. As previously mentioned, Hakimullah Mehsud was a Taliban subcommander who had gained fame by attacking U.S. and Coalition supply convoys traveling through the Khyber Agency to Afghanistan. The new Pakistani Taliban chief lost no time in declaring his "love and affection" for America's number-one enemy, Osama bin Laden, and promised swift revenge on the CIA for the death of his friend and predecessor, Baitullah Mehsud.[27]

Hakimullah ended his message to the Americans by criticizing them for imprisoning Muslims in Abu Ghraib, Iraq, and Guantánamo Bay, Cuba, and predicted, "If America continues to attack the innocent people of the tribal areas then we are forced to attack America." Then he added, "We will make new plans to attack them. You prepare for jihad and this is the time of jihad."[28] In essence Hakimullah was invoking the ancient Pashtun tribal code of *badal*, which calls for eye for an eye revenge against one's enemies, regardless of the cost.

Events were to show that Hakimullah's call for vengeance against the CIA murderers of his former mentor, Baitullah Mehsud, and countless other Taliban and Pashtun tribesmen were no mere words. Hakimullah eventually kept his promise by killing a CIA station chief linked to the drone attacks and seven of her fellow officers. But the feud did not end there. Hakimullah's commander who was in charge of the revenge attack on the CIA would later be killed by a drone. Hakimullah himself was later reported wounded in the legs and abdomen, but not killed, in a subsequent drone strike.[29]

Thus the cycle of violence in the rugged mountains of the Afghan-Pakistani border perpetuated itself in a way that many previous conquering states and empires had experienced over the centuries. How many other Pashtuns had similarly declared *badal* on the United States as a result of the bloody drone campaign against Baitullah Mehsud no one knew. How many innocents had been killed in the numerous strikes on him and his followers before he was finally assassinated? Were the Americans making more enemies than they could kill, or were they simply using the most advanced means at their disposal to eradicate dangerous men who were committed to causing future slaughter and terrorism? No one seemed to know the answers to these important questions.

While the debate on these issues has been driven by extremes (the arguments that on the one hand, "drones make more enemies than they kill" and, on the other, "they are an unprecedented means for killing al Qaeda and Taliban members"), this book will try to find a middle ground. It will do so by analyzing the wider issues involved in the drone attacks, such as the unique history of the Pashtun tribal areas, Pakistani relations with the Taliban and the United States, the development of the armed drones, Pakistani reactions to the drone strikes, and Taliban and al Qaeda responses. By looking at all aspects of the issue, one can construct a three-dimensional picture of this murky assassination campaign that is still not fully understood even by those carrying it out or those suffering from it.

Before these issues can be explored the reader must first, however, make a crucial background journey into the missing history of the remote Pashtun territory where the drone strikes have been carried out, the FATA. It is only by understanding the culture and history of this autonomous land that one can understand the ebb and flow of the drone war that is taking place on and above it.

2

A History of the Pashtun Tribal Lands of Pakistan

*From a historical perspective, this ignorance about the enemy
makes the war on terror unique. Rarely have so many resources
been deployed on the basis of such a vague understanding
about who the enemy is and how it functions.*
—**Thomas Heggerhammer,** *Times* **(London), April 2, 2008**

*Afghanistan is a war of attrition, because Pakistan
provides a sanctuary for the enemy.*
—**Bing West,** *The Wrong War*

T he Pashtuns who live in the autonomous tribal zone targeted by the
drones are an Aryan–East Iranian people dominated by the tribal
code of Pashtunwali who are often called the Afghans or Pathans.
The Afghan-Pashtuns created a state in the mid-1700s that included modern-day Afghanistan and the Pashtun tribal lands of northwestern Pakistan.
When the colonial British advanced from India to the borders of their state
in the 1800s, the Pashtun hill men began to raid British India. This set off a
series of border skirmishes between the expansionist British and the unruly
Afghan-Pashtuns that led to the British conquest of a sizable chunk of
Pashtun territory, which was ultimately added to their vast Indian empire.
The Pashtuns were thus divided between Afghanistan and British India (portions of the latter became Pakistan in 1947).

THE FATA, 1947–1998

The British then carved the Pashtun territory they had annexed into India into two zones known as the FATA and the North-West Frontier Province. The North-West Frontier Province was less hilly and easier to tame than the FATA, and it was often described as "settled." This area subsequently became a regular province of British India and later the newly independent country of Pakistan. But the Pashtuns in the hills of the FATA were more unsettled, and they revolted against their British masters on many occasions. For this reason they were never fully included in the British state the way the provinces of Sindh, Punjab, Baluchistan, and the North-West Frontier were. A state of perpetual low-level rebellion that often turned into outright war existed in the FATA for most of its history.

In 1947 the newly independent state of Pakistan thus inherited its western borders, defined by the FATA, from the British, and it has kept the system in place to this day. The Pakistani army never entered the FATA, and the government rarely meddled in this autonomous tribal region except to meet with their intermediaries with the tribes, the political agents. The agents ruled the tribesmen through the *maliks*.

The FATA has thus remained the wildest and most undeveloped part of Pakistan. It was, and still is in many ways, a world unto itself. It has a high poverty rate, low levels of literacy, and few schools and roads, and its population is deeply conservative in religious and tribal terms. Whereas the Sindhis and Punjabis who dominate Pakistan tend to be relaxed Barelvi Sufi Muslims who are strongly influenced by Indian culture and all that it entails (from Bollywood to clothing styles), the Pashtuns of the FATA are much more conservative and have been drawn to the fundamentalist Deobandi branch of Islam.

The FATA might have remained an obscure, Massachusetts-sized conservative backwater of 3 million people had it not been for the December 25, 1979, invasion of neighboring Afghanistan by the Soviet Union. Owing to its strategic location on the Afghan-Pakistani frontier, the FATA quickly became a springboard for cross-border Pashtun jihad against the "godless" Soviet invaders. When the Soviets invaded Afghanistan, they began to fight with Afghan-based Pashtun mujahideen (holy warrior) rebels who declared a holy war on the Russian unbelievers. Such Afghan-Pashtun mujahideen leaders as the fanatical Gulbuddin Hekmatyar and Jalaludin Haqqani used Pakistan's FATA and neighboring North-West Frontier Province as rear-area staging grounds for carrying out cross-border strikes on the Soviet invaders. Wounded or weary mujahideen trekked over the mountains into Pakistan to regroup and replenish their supplies in the FATA or the North-West Frontier capital of Peshawar. The mujahideen rebels' weapons caches were safe from the enemy in the FATA, and their guerrillas could evade Soviet offensives by fleeing across the border.

Not surprisingly, the Soviets, who were losing up to two thousand men a year to the mujahideen, responded much as the Americans had in Vietnam—by trying to bomb their enemies' cross-border sanctuaries. According to the Pakistanis, the Soviet air force entered Pakistani airspace more than two thousand times in pursuit of mujahideen, often bombing deep into sovereign Pakistani territory.[1] On these bombing raids the Soviet MiG and Sukhoi fighter-bombers used unguided "dumb" bombs, which were extremely imprecise, and this led to the deaths of scores of Pakistani civilians. The Pakistanis responded aggressively to the violations of their airspace and used their new American-supplied F-16 fighter jets to shoot down eight Soviet aircraft.[2] (Such robust defense of their airspace contrasts drastically with the Pakistanis' current policy of allowing the United States to deploy its slow-moving drones against the Taliban in the same region.)

This being the period of the Cold War, the American CIA quickly got involved in the act as well. In the FATA the mujahideen "freedom fighters" were armed by CIA operatives who gave them cash, radios, Blowpipe weapons, and Stinger antiaircraft missiles to fight their Russian adversaries. The CIA worked closely with Pakistan's Islamist president Mohammad Zia-ul-Haq and his ISI to arm the 200,000 mujahideen so that they could "bleed" the Soviet occupation forces in Afghanistan.

Although it is not true, as many have suggested, that the United States trained and equipped Arab jihad volunteers such as bin Laden, it is true that the United States trained and equipped Hekmatyar and Haqqani. This policy was to later backfire and lead to the so-called blowback effect. Today, Haqqani is the most effective Taliban leader in the eastern Afghanistan provinces of Khost, Paktia, and Paktika, while Hekmatyar is a Taliban-linked insurgent commander in northern Afghanistan. In a case of déjà vu, both of these Pashtun mujahideen-turned-pro-Taliban-insurgents continue to use the FATA as a sanctuary, but this time it is against their former sponsors in the 1980s jihad, the Americans.

They say that war makes strange bedfellows, and this was never more true than with the Americans and President Zia-ul-Haq, a dictatorial, Islamist general who had seized power in Pakistan in 1977 from the democratically elected president of the country, Zulfikar Bhutto (the father of Benazir Bhutto). After the coup, the elder Bhutto was put on trial and executed. When the Soviets were finally driven from Afghanistan in 1989, the Americans began to see the Pakistani leader's flaws in a clearer light.

Differences between the two anti-Soviet allies were heightened in 1990, when the United States could not confirm that post-Zia Pakistan was not developing a nuclear weapon. The Americans were deeply concerned about the spread of nuclear weapons to this unstable country and enacted the Pressler Amendment to enforce sanctions on Pakistan for trying to develop them. This led Washington to withhold $1.2 billion worth of military equipment that the Pakistanis had already paid for. Among the most important weapons withheld by the Americans was a delivery of F-16 fighter jets.[3]

The Pakistanis were of course furious, and the military and ISI felt they were being used by the Americans. At this time Pakistani officers who had been traveling to U.S. military academies ceased coming to America for training. A new generation of Pakistani military officers would come of age in the 1990s distrusting the Americans and believing in Islamist causes such as the jihad against the Indians in the contested province of Kashmir. These same officers would later be asked to join with the Americans in waging war against the Taliban and al Qaeda after 9/11. One Pakistani officer summed up this generation's views of America when he said, "You used us, and then you dumped us. Pakistanis are convinced you are going to do it again."[4] There was bound to be mistrust between the two allies who had parted ways during the 1990s under a cloud of mutual suspicion.

There were deeper differences between the two nations as well. Pakistan had become an increasingly fundamentalist place under President Zia-ul-Haq, who died in 1989. Zia-ul-Haq had built thousands of madrassas (religious seminaries) to increase Islamic fervor in his country and unite its various peoples under the umbrella of Islamism. Many of these Saudi-funded madrassas were built in the FATA and served as orphanages for young Afghan refugee boys. The students known as Taliban (Taliban is the plural of the Pashto and Persian word for "student"—*talib*) who studied in the madrassas in the late 1980s and 1990s were even more conservative than the already conservative Pashtuns.

AFGHANISTAN, 1988–2001

By the early 1990s United States had come to see its former mujahideen proxies as a terrorist threat as well. During the first Gulf War with Iraq, the former CIA-backed Afghan mujahideen leader Gulbuddin Hekmatyar supported Saddam Hussein in his fight against the Americans.[5] Then, in 1993, a Pakistani named Ramzi Yousef, who had trained in an Afghan mujahideen camp, tried to blow up the World Trade Center. At this time the United States threatened to designate Pakistan as a state sponsor of terrorism. Additional sanctions were imposed on Pakistan when it bought banned missile technology from Communist China.

In the meantime, post-Soviet Afghanistan had descended into an all-out civil war between the various mujahideen factions. The worst offender in the war was the Pakistanis' candidate to rule in Kabul, Hekmatyar. He mercilessly shelled the Afghan capital, killing thousands of civilians in his effort to take the city. The siege led to a collapse of authority and to chaos in the Pashtun south. There the ex-mujahideen attacked civilians, set up checkpoints to prey on the common people, and raped women. The southern province of Kandahar in particular was consumed by violence as predatory Pashtun mujahideen bandits made it all but impossible to travel.

The violence in the Afghan south disturbed many Afghan Pashtuns, none more so than a former mujahideen leader–turned–mullah (cleric) named Omar. In 1994 Mullah Omar led a group of religious students in attacking a local mujahideen warlord who had recently attacked a traveling family, killed its men, and turned its women into sex slaves. Mullah Omar's Taliban overran the warlord's checkpoint and famously hung him from the barrel of a tank.

When word of Omar's success spread, his Taliban movement began to snowball. Although there had always been Taliban in the Pashtun tribal areas, the name began to take on a new meaning. It came to mean black-turbaned vigilantes who were fighting to end the chaos and violence of the ex-mujahideen. One by one the Pashtun mujahideen of the south were either killed, disarmed, or grew their beards long and joined the Taliban movement.

By 1995 the Taliban had begun to spread from beyond Kandahar into the western lands of the ethnic Tajiks and into the lands of the northeastern Pashtuns known as the Ghilzais. The most powerful Ghilzai Pashtun mujahideen leader in this area was the previously mentioned Jalaludin Haqqani. Haqqani made the decision to join the Taliban and brought his troops over with him. As a mark of respect, Haqqani was then made Taliban governor of Paktia Province and minister of tribes and frontiers.

By this time the Pakistani ISI had begun to fully bet on the Taliban in their quest to conquer all Afghanistan. There were many reasons for this realpolitik decision. On the one hand, Taliban members were first and foremost Islamists. They did not organize officially on the basis of their Pashtun ethnic identity. This was a huge relief to the Pakistanis who had long feared what would happen if the Pashtuns living in Pakistan and those living in Afghanistan united in an effort to create a larger "Pashtunistan." This Pashtun state could be created only by slicing land from Pakistan's northwest and adding it to Afghanistan. The Pakistanis, who had already lost East Pakistan (i.e., Bangladesh), were not going to lose the Pashtuns who made up 15 percent of their state to Afghanistan.[6]

For this reason, the Pakistanis actively supported the Afghan Taliban in their battle for supremacy with a coalition of Uzbek, Tajik, and Hazara ethnic opponents known as the Northern Alliance. While much of this support for the Taliban was ad hoc and consisted of thousands of Pakistani madrassa students (mainly Pashtuns) who came to fight for the Taliban in the summer, the Pakistani government also directly supported the Taliban with volunteer soldiers.

It did not bother the Pakistanis that on many levels the people of Afghanistan had begun to suffer terribly under the Taliban. The Taliban members had originally been greeted as Robin Hood–style heroes when they conquered the Pashtun mujahideen warlords of the south, but they had become more and more puritanical as they spread out into the Uzbek, Tajik, and Hazara lands. Among other things, they horribly oppressed half the population of Afghanistan, the women, by denying them the right to work.

The Taliban also arrested women caught outside without a male relative, cut the fingers off women who wore nail polish, and enforced the Pashtun burqa in areas like cosmopolitan Kabul and the Hazara lands where it had not previously been enforced.

The men also suffered under Taliban rule. They had to grow full-length beards and appear in mosque at prayer time. The Taliban closed movie theaters; destroyed TVs, which were deemed "satanic devices"; executed homosexuals, elopers, and adulterers; closed girls' schools; and banned almost all forms of recreation (for example, music, dancing, kite flying, and singing). Under the Taliban, Afghanistan was essentially converted into a grim religious prison camp dominated by Pashtun zealots.

Such issues did not originally bother Washington policymakers (although the Americans played no role whatsoever in the creation and rise to power of the Taliban as some have suggested). For all their distaste of the Taliban's social policies, the distant Americans were content to see them consolidate power and create stability in this war-torn land.

But Washington eventually grew hostile to Afghanistan's new Taliban masters when they began to play host to an international terrorist organization known as al Qaeda. Although the Taliban leader Mullah Omar tried to control his Arab terrorist guest, Osama bin Laden, and put an end to his angry calls for total jihad against the American "Far Enemy," bin Laden proved to be uncontrollable. In August 1998 bin Laden's al Qaeda terrorists blew up U.S. embassies in Kenya and Tanzania killing hundreds of Africans and a few Americans. President Bill Clinton showed America's resolve to kill the terrorists responsible for the slaughter by launching a wave of cruise missiles against bin Laden's bases, found primarily in the territory of his old friend from the 1980s Afghan jihad, Jalaludin Haqqani (incidentally, the distrusted Pakistanis were warned only at the last minute that a wave of cruise missiles would be crossing their territory to hit bin Laden's terrorist camps in eastern Afghanistan).

But the strikes failed to kill their intended target. Far from killing bin Laden, the strikes only infuriated Mullah Omar, who vowed to protect his own people and Arab terrorist guests from the American "infidels."

President Bill Clinton's lawyers at the time concluded that the United States could legitimately kill bin Laden and his lieutenants, despite a previous presidential ban on assassinations. The lawyers concluded that attempts to kill bin Laden were defensible as "acts of war" or as "national self-defense" under both American and international law.[7] Clinton subsequently signed a presidential finding and issued secret orders allowing the CIA to assassinate bin Laden.[8] These findings, and a later one by President George W. Bush, would give the CIA the authority to carry out the drone assassination campaign on al Qaeda and the Taliban after 9/11.

Although it was not discussed much at the time, Clinton's presidential finding was a groundbreaking development. The ban on government-sponsored assassinations went all the way back to President Gerald Ford. During his tenure in the White House, Ford had discovered that under Presidents Eisenhower, Kennedy, Johnson, and Nixon, the CIA had tried to murder a grand total of eight foreign leaders, including Fidel Castro of Cuba. In 1976 President Ford authorized Executive Order 12333, forbidding anyone in the U.S. government from engaging in such assassinations.[9]

In the years following the failure of the assassination attempt on bin Laden, the United States scrambled to develop new means for carrying out more precise airborne raids on enemy targets. By the time Bush took over from Clinton as president of the United States, that mission was well on its way. As bin Laden prepared for his greatest terrorist outrage, the so-called Holy Tuesday attacks of 9/11, the ultrasecret Defense Advanced Research Projects Agency (DARPA) was busy working on a new advanced weapon that would give the CIA the ability to kill enemies with unparalleled precision. This was to lead to one of the most important developments in air warfare since the first pilots brought guns and bombs into their aircraft during World War I.

3

Enter the Predator

Every so often in the history of war, a new weapon comes along that
fundamentally rewrites the rules of battle.
—Lara Logan, discussing the advent of drones on *60 Minutes*

T he August 2008 Predator missile strike that killed Pakistan's most
wanted man, Taliban leader Baitullah Mehsud, did not come about
overnight. The story of the development of the CIA's top-secret
Predator-Reaper program is one of the murkiest chapters in the war on terrorism. The story of how a clumsy, propeller-pushed, remote-control plane
went from the drawing boards to becoming the most effective, high-tech
assassination tool in history has been shrouded in mystery.

VIETNAM, THE BALKANS, AND IRAQ, 1970–2000

The origins of the program actually lay in the Vietnam War era. This was the
time of the CIA's infamous Phoenix Program, which saw the agency's antiterrorism teams assassinate thousands of communist infiltrators and terrorists.
During the Vietnam War the United States began using remote-controlled
drones known as Lightning Bugs to fight the enemy. The CIA and Air Force
used the jet-propelled Lightning Bugs to carry out high-altitude photo reconnaissance missions against the Viet Cong and North Vietnamese.[1]

The real predecessors to the Predator, however, were the UAVs known
as the Gnat and Amber. The Amber drone was first built by a U.S.-based
company owned by a former Israeli air force designer, Abraham Karem,

called Leading Systems. It was developed in the late 1980s and then reconstituted as the Gnat, a drone that was similar to the modern Predator in shape and configuration. The Predator drone used in Pakistan with such deadly effect after 2004 was developed from the Gnat and made its debut flight in June 1994. This initial Predator, built by General Atomics Aeronautical Systems in San Diego, was a spy/surveillance aircraft and was not armed.

The timing of the Predator's development was most serendipitous for the CIA because the U.S. military had recently become involved in a war in the Balkans. The United States desperately needed a new reconnaissance aircraft to spy on the enemy in this complex civil war. Interestingly enough, the Predators (four initially) were deployed in July 1995 to spy on the Christian Serbs who had slaughtered thousands of Bosnian Muslims in the worst case of genocide in Europe since the Nazis.[2] The United States intervened to fight on behalf of the Muslims to prevent further genocide and began an aerial campaign against Republika Srbska Serbian troops known as Operation Deliberate Force. To help U.S. attack aircraft spot Serbian targets, the CIA Predators, which were based in Gjader, Albania, made recon flights over Bosnia in an operation known as Nomad Vigil.[3] The actual pilots for the aircraft were based in trailers in Indian Springs Air Force Base, Nevada (renamed Creech Air Force Base in 2005) and belonged to the Eleventh Reconnaissance Squadron. The CIA's dream of using remote-control planes to collect data and intelligence from the skies had finally come true.[4] Author and journalist Steve Coll describes the revolutionary development as follows: "In the first flights over Bosnia the CIA linked its Langley headquarters to the pilots' van. Woolsey [the CIA head] emailed a pilot as he watched video images relayed to [CIA Headquarters] Virginia. 'I'd say What direction for Mostar? . . . Is that the river? . . .' Woolsey recalled. 'And he'd say Yeah. Do you want to look at the bridge? . . . Let's zoom further, it looks like he has a big funny hat on.'"[5]

These early surveillance drones were a huge improvement on the existing surveillance option: orbiting satellites. Unlike spy satellites whose views were blocked by clouds, the drones could fly under the cloud cover to monitor their targets. Once they found their target, they could follow it for hours at a time, unlike satellites, which flew over their designated targets only when their preexisting orbits took them there. And satellites, while useful in film-ing static targets, were less capable of filming small moving targets, such as vehicles or humans.

Drones were also far superior to "fire and forget" cruise missiles, which were usually launched from offshore vessels and took considerably more time to reach their destination. The situation in the targeted area could change dramatically while the less accurate cruise missiles made their way to their preprogrammed target. When finally armed, drones by contrast could fire in live time based on their pilots' reaction to current information gained from tracking fluid targets.

For all their revolutionary advances, however, these experimental Predators were not yet equipped with radar systems that would allow them to see through cloud cover. They were finally fitted with radars that could allow them to see through fog and clouds in 1997, demonstrating that the Air Force was still perfecting the early drones.

In January 1999 Predators were flown to the Persian Gulf and used to spy on Saddam Hussein's Iraq as part of Operation Southern Watch.[6] In the spring of that year Predators were also used against Serbian Christian forces that were once again engaged in genocidal assaults on Muslims, this time the Kosovar Albanians. By all accounts, the Predator "eye in the sky" gave the Americans unprecedented access to Serbian troop movements and facili-tated the accuracy of U.S. bomb strikes.

At this time the Predators made another technical leap when laser des-ignators and range finders were added to the censor balls on their "chins." This meant that the Predator could lase a target and a loitering manned fighter jet could then use a laser-designated bomb to precisely destroy it.[7] The ball under the Predator's chin was one of its most expensive features and also came to contain two television cameras, including an infrared camera for seeing targets on the ground at night and the previously men-tioned radar, which could see through clouds and dust. Pilots watching the screens back in bases outside of Las Vegas, Nevada, reported that this allowed them to see a license plate from more than two miles away in the air.[8] This was a capability the Predator would need for its next mission in Central Asia.

AFGHANISTAN, 2000–2001

By 2000 America's focus had transferred from the war-torn Balkans to the Taliban-controlled Islamic Emirate of Afghanistan. It was here that Arab terrorists belonging to bin Laden's al Qaeda were plotting further terrorist attacks on the United States. In 2000 bin Laden's agents set off a bomb next to the destroyer USS *Cole* in Yemen, killing seventeen sailors. The same year an al Qaeda agent named Ahmed Ressam attempted to infiltrate America to set off a bomb at Los Angeles's LAX airport.

While many Americans, who lived far from al Qaeda's targets in eastern Africa and Arabia, remained blissfully unaware of the danger posed by this terrorist group, many in the CIA saw the organization as the greatest threat to the continental United States. The United States had by this time established a Bin Laden Unit at the CIA; in fact, some CIA personnel called this unit the "Manson Family" for its members' obsession with the little-known terrorist bin Laden.

In the spring of 2000 the United States gained permission from the dictator of Uzbekistan, President Islam Karimov, to fly Predator surveillance drones out of his country over southeastern Afghanistan to try tracking bin Laden. As mentioned previously, the Clinton administration had already overridden President Ford's earlier directive against carrying out assassinations. It was hoped that the Air Force–piloted Predator could help the CIA track down bin Laden, who was known to live in a series of compounds in the Pashtun lands of southern and eastern Afghanistan. The drones, it was theorized, might then be able to direct a cruise missile strike against bin Laden from a submarine or cruiser operating in the Indian Ocean. This joint Pentagon-CIA surveillance operation was to be known as "Afghan Eyes" and was headed up by White House counterterrorism adviser Richard Clarke, CIA counterterrorism chief Cofer Black, and Charles Allen, head of the CIA's intelligence-gathering operations.[9] Clarke in particular was worried about al Qaeda's ability to hit the U.S. mainland in the months and years before 9/11 and was interested in any tool that might help him prevent such an event.

In the fall of 2000 a drone monitoring a known bin Laden compound in the Afghan south near Kandahar at a place called Tarnak Farms sent back live video feed to a screen at the CIA's Counterterrorism Center in northern Virginia. The extraordinary video was of a tall man (bin Laden was six foot six), "with a physical and operational signature fitting Bin Laden," wearing white robes and talking to ten figures who were paying him respect. Those watching the screen in America were stunned and later said, "It was probably Bin Laden himself."[10] In fact the drone may have spotted bin Laden on as many as three separate occasions.

There was little the Predator could do because such drones were not armed at the time, and the CIA subsequently lost him. Those CIA operatives who saw the images of "the man in white" on their screen in America were frustrated. As one later put it, "If we had developed the ability to perform a Predator-style targeted killing before 2000, we might have been able to prevent 9/11."[11]

Around this time the Pentagon and the CIA began to seriously contemplate arming the Predator and transforming it from a "sensor" into a "shooter." As one general involved in the development by DARPA put it, "If the drones were equipped with laser-guided targeting systems and weapons, then the whole cycle—from finding a target and analyzing it to attacking and destroying the target and analyzing the results—could be carried out by one aircraft."[12] The drones would now be part of the "kill chain," not just unarmed spotters for armed aircraft or cruise missiles.

The two men in charge of the operation to arm the Predator were Gen. John Jumper, the U.S. Air Force chief of staff from 2001 to 2005, and James Clark, the Pentagon's chief of staff of intelligence.[13] Both men understood that they had been given a difficult task. The Predator was a lightweight, flimsy reconnaissance aircraft, and many doubted it could carry heavy munitions on its gliderlike wings. According to the experts, the Predator could only carry a missile and launch rails that weighed less than 175 pounds. This excluded most munitions.

After considerable searching, the Air Force hit upon an ideal lightweight weapon that fit this criteria—the AGM-114 Hellfire missile. The Hellfire is a hundred pound, antiarmor, air-to-surface missile. It had been designed to be fired primarily at tanks by attack helicopters. Before a Predator could use it, the Hellfire had to be reconfigured because it tended to penetrate nonarmored targets and explode in the ground beneath them. The U.S. Army solved the problem by fixing the Predator's Hellfire missiles with metal sleeves that caused deadly shrapnel and fragmentation when they exploded.[14]

The overly penetrative nature of the Hellfire was not the only worry the Air Force had about the missile. General Jumper feared that when it was fired, the powerful Hellfire would break off the Predator's fragile wings. Everyone involved waited in anticipation for the test firing of the Predator's first Hellfire missile on a test flight in Indian Springs, Nevada. There, on February 16, 2001, Predator number 3034 took off on a test flight and successfully fired its Hellfire missile at a tank. The RQ-1 Predator soon thereafter lost its *R* (reconnaissance) designation and was renamed the MQ-1 (the *M* for multimission). It was a revolutionary moment in the history of aerial warfare. The unmanned reconnaissance drone had become a killer.

There was no doubt about who the remote-control killer's first target would be. In a display of its future intentions, a Predator was subsequently used to fire a Hellfire missile at a mockup clay compound in Nevada built to resemble a typical house in Afghanistan.[15] As the *Washington Post* put it, "The Bush administration now had in its hands what one participant called 'the holy grail' of a three-year quest by the U.S. government—a tool that could kill bin Laden within minutes of finding him."[16]

The $4.5 million Predator could fly 420 miles, then circle over a target for up to thirty hours, and feed real-time video through ten simultaneous streams to controllers in ten different locations. This, of course, made it ideal for finding bin Laden. The Predator also carried sensors that intercepted electronic signals and listened in on phone conversations. It was more than just a weapon; it was an eye and ear in the sky.

Richard Clarke, who continued as the White House's chief counterterrorism adviser under the new president, George Bush, advised the new national security adviser, Condoleezza Rice, to focus on Afghanistan, where bin Laden was hiding, and not on Iraq and Saddam Hussein. He also stressed the importance of using the newly armed Predator drone to track down bin Laden and assassinate him.[17]

But CIA head George Tenet had serious qualms about the new killing technology and the ethics and legality behind its use. The consensus in the CIA was that "aircraft firing weapons was the province of the military."[18] According to one former intelligence officer, "There was also a lot of reluctance at Langley to get into a lethal program like this."[19]

The branch of the military that would be asked to fly the drones, the Air Force, was similarly disinclined to take charge of them. Steve Coll writes, "The Air Force was not interested in commanding such an awkward, unproven weapon. Air Force doctrine and experience argued for the use of fully tested bombers and cruise missiles, even when the targets were lone terrorists. The Air Force was not yet ready to begin flying or commanding remote control planes."[20]

According to Coll, "James Pavitt as the Director of Operations at CIA was also worried about the unintended consequences should the CIA suddenly move back into the business of running lethal operations against targeted individuals—assassination in the common usage."[21] For all its potential, neither the Air Force nor the CIA was inclined to embrace the new remote-control technology or its potential role as a terrorist killer on the eve of 9/11. Far from being trigger happy, Tenet wanted the government to have its "eyes wide open" to the ramifications of using the drones to assassinate terrorists.[22] He was said to have been "appalled" at the question as to who should "pull the trigger" on bin Laden or other terrorists and did not seem to feel that he had the jurisdiction to do so.[23] In his autobiography he asked, "How would the government explain it if Arab terrorists in Afghanistan suddenly started being blown up?"[24]

The American government had previously been critical of the Israeli policy of assassinating its Hamas and Hezbollah terrorist enemies. Ironically, as recently as July 2001 the U.S. ambassador to Israel, Martin Indyk, had stated, "The United States government is very clearly on record as against targeted assassinations. . . . They are extrajudicial killings, and we do not support that."[25]

Now the CIA was potentially being tasked to do the same thing as the Israelis, only it would be done via an unexplored new technological device whose ethics and morality were not fully understood. Capturing the CIA's unease, Tenet stated, "This was new ground." He asked, What would be the chain of command should the Predator be used, who would take the shots, and were America's leaders comfortable with the CIA doing this killing outside the military's normal command and control?[26]

As a result of jurisdictional squabbles over who would pay for and fly the drones and moral qualms about their use, discussion on deploying the

Predator to kill bin Laden was shelved in a September 4, 2001, meeting involving key government officials.[27] Just days before 9/11, "terrorism was not at the top of the priority list of the new Bush administration."[28] With no real sense of urgency in the air, talk of what to do with the Predator was put off to a later date.

4

Operation Enduring Freedom

The gloves are off. Lethal operations that were unthinkable
pre–September 11 are now underway.
—**Senior White House official after the 9/11 attacks**

War is the mother of invention, and the unexpected destruction of 9/11 led to a global war that was to see tremendous developments in America's killing technology. Although the Bush administration had been obsessed with Baathist Iraq since it had come to power in 2000, the death of almost three thousand people on 9/11 abruptly diverted the White House's attention to Central Asia. The White House was now focused on the clear danger to American lives emanating from the previously ignored Islamic Emirate of Afghanistan. Clearly the president had to respond to the unprecedented destruction and move to defend his people. But how?

For his part, bin Laden was confident that the United States would react to the 9/11 attacks as it had after the 1998 embassy bombings in Africa, that is, by launching punitive cruise missile strikes. But it was clear to the White House and Pentagon that something more drastic was necessary. Al Qaeda's sanctuary in Afghanistan needed to be totally destroyed if America was to be made safe again. This meant convincing the Taliban host regime to arrest the hundreds, if not thousands, of Arab jihadists in their country.

But the Taliban reacted to the stunning news from distant North America by panicking and denying their Arab guests' guilt. When confronted with the news of the Taliban's intransigence, the Bush administration had no recourse but to move against the country. Richard Armitage, the deputy secretary of state, summed up America's position as follows: "We told the Taliban in no uncertain terms that if this happened, it's their ass. No difference between the Taliban and Al Qaeda now. They both go down."[1]

Infuriated by Mullah Omar's decision to stand by al Qaeda, President Bush ordered his top general to "rain holy hell" on the Taliban.[2] As for bin Laden, Cofer Black, the head of the CIA's Counterterrorism Center who had been active in pushing for the arming of the Predator drone, was more blunt. Black ordered his Special Activities operatives to "capture Bin Laden, kill him, and bring me his head back in a box on dry ice."[3]

President Bush showed his newfound resolve to tackle al Qaeda, a group his administration had not been overly interested in prior to 9/11, with his vow to capture bin Laden "dead or alive." National Security Adviser Condoleezza Rice subsequently hinted at what was to come when she said, "We're in a new kind of war, and we've made it very clear that this new kind of war will be fought on different battlefields."[4] Bush himself added, "Traditional concepts of deterrence will not work against a terrorist enemy whose avowed tactics are wanton destruction and the targeting of innocents."[5] A senior U.S. official further demonstrated how far the administration had come from criticizing Israel for targeted assassinations when he stated, "The gloves are off. Lethal operations that were unthinkable pre–September 11 are now underway."[6] And another U.S. official advocated using "all the weapons at our disposal" to target bin Laden and his followers.[7] Thus the foundations were laid for a veritable revolution in counterterrorism (and ultimately counterinsurgency) even before the rubble from the World Trade Centers had begun to be cleared.

Drawing on the presidential finding written for Bill Clinton in 1998, which allowed the CIA to assassinate bin Laden, the Bush administration concluded that executive orders banning assassination did not prevent the president from lawfully targeting terrorists for assassination. President Bush subsequently signed his own finding to that effect. At the time the *Washington Post* reported, "The CIA is reluctant to accept a broad grant of authority to hunt and kill U.S. enemies at its discretion, knowledgeable sources said. But the agency is willing and believes itself able to take the lives of terrorists designated by the president."[8]

On September 14, 2001, the Senate voted 98–0 and the House voted 420–1 to pass the Authorization for Use of Military Force (AUMF) as a joint resolution that "authorized the use of United States Armed Forces against those responsible for the recent attacks launched against the United States."[9] Bush and later Obama would argue that this gave them the mandate to hunt al Qaeda terrorists wherever they were found. Critics would point out the AUMF gave the military, not the civilian CIA, the authorization to go after terrorists, but for the Bush White House, this was mere semantics.[10] It was time to launch a full-scale preemptive campaign of deterrence that would translate to a hunt for terrorists who were plotting the deaths of more Americans.

As the events were taking place in the fall of 2001, the U.S. secretary of state, Colin Powell, and his deputy, Richard Armitage, were engaged in a different mission that was to have equal ramifications for the upcoming war against the Taliban and al Qaeda. They were trying to convince the Pakistanis to turn on the Taliban proxies that they had nurtured since 1994 and join the United States in the newly declared global war on terror (GWOT) against them and al Qaeda. Talks on this delicate issue were facilitated by a visit to Washington, DC, by the head of the Pakistani ISI, Gen. Mahmoud Ahmed, on the fateful day September 11, 2001. He saw for himself the way the stupendous destruction and loss of lives on that day had transformed the American officials around him. He reported this transformation back to his leader, Pakistani president Pervez Musharraf. It was clear from Ahmed's report that the slaughter of three thousand people in Manhattan, Virginia, and Pennsylvania had changed the Americans much as the attack on Pearl Harbor had in 1941. Ahmed detected a newfound determination to wage war against al Qaeda by all means possible.

Although President Musharraf and the Pakistani leadership were distrustful of the Americans, who had previously sanctioned them for building nuclear weapons under the Pressler Amendment, they were given an offer they could not refuse. If they agreed to join the newly belligerent Americans, the United States would provide them with billions of dollars in aid known as Coalition Support Funds. Washington would also lift the previously imposed sanctions and help arm the Pakistanis. If they did not join the Americans, they would be declared a state sponsor of terrorism, and economic sanctions would be leveled against them.

President Musharraf put his finger to the wind and sensed that it was blowing against the doomed Taliban regime and its terrorist allies. Armitage had told the Pakistanis, "You are either 100 percent with us or 100 percent against us—there is no grey area."[11] The world's lone remaining superpower and its powerful North Atlantic Treaty Organization (NATO) allies were clearly going to go to war against the mullahs in Afghanistan. Musharraf's money was on the United Nations (UN)–sanctioned American-led NATO alliance. For this reason he chose to drastically alter his country's foreign policy vis-à-vis its Taliban allies. Pakistan was now "with" the Americans and would become the United States' most important strategic partner in the upcoming war in Afghanistan.

Such a drastic redirection of foreign policy was not easy for Musharraf to carry out. After a heated meeting with his top generals and the firing of those who wanted to stand by their Taliban surrogates, Musharraf finally agreed to one of the Americans' most important requests: the right of "blanket over-flight and landing."[12] Among the Pakistanis' first actions in their new alliance with the distrusted Americans was to give the U.S. military the right to use several intermediate staging bases located in the province of Baluchistan, at Jacobabad, Pasni, Dalbandin, and Shamsi. Later, two of these air bases, at Jacobabad and Shamsi, would be shown to have been used by Predator drones. There were also rumors that U.S. drones were flying from a Pakistani base at Ghazi.

It can be surmised that the Air Force and CIA flew drones into these bases in time for the October 2001 attack on the Taliban of Afghanistan, that is, the beginning of Operation Enduring Freedom. Only by understanding this important American-led operation and the course of the war on the Taliban and al Qaeda can one understand the CIA's subsequent drone air campaign against the Taliban in the FATA. Events in this little-studied operation led to the development of a terrorist-insurgent sanctuary in Pakistan's autonomous tribal agencies that continues unabated to this day. The existence of these safe havens in the Pakistani tribal zones led to the commencement of the CIA's post-2004 drone war in Pakistan.

Operation Enduring Freedom began on October 7, 2001, with an airborne campaign. Having achieved air supremacy in just a few days the United States and its Coalition allies then began to bomb Taliban positions throughout Afghanistan.[13]

At this time a legendary anti-Taliban Pashtun mujahideen leader named Abdul Haq sneaked into Afghanistan to try leading a rebellion against the Taliban in the Pashtun south. Unfortunately, his small group of followers was ambushed by the Taliban and took heavy losses. As the Taliban closed in on him, Abdul Haq desperately radioed U.S. Central Command for help. The Americans responded by sending an armed Predator drone to attack the Taliban. But before the slow-moving drone could reach Haq, the Taliban overwhelmed him and captured him and his followers. The Predator arrived subsequently and fired on a Taliban convoy at the scene of the capture, but it was too late to save the doomed leader.[14] Haq was tortured and then hung by the Taliban.

America's air war was not going well on other fronts either. As the U.S. Air Force, Navy, and Marines bombed the country, the Taliban simply dug in and appeared to be prepared to ride out the U.S.-led Coalition's deadly air strikes. But the Coalition had plans to move on the ground as well. The U.S. Army's Fifth Special Forces Group planned on using the anti-Taliban Northern Alliance as a proxy ground force. These Uzbek, Tajik, and Hazara opposition fighters, who had been pushed into small mountain sanctuaries in the northeast and central Hindu Kush Mountains, were all too happy to join the Special Forces in fighting their Pashtun-Taliban blood enemies.

Because the overlord of the Northern Alliance opposition, Massoud the Lion of Panjsher, had been killed by al Qaeda suicide bombers on September 9, 2001, the larger-than-life Uzbek leader Gen. Abdul Rashid Dostum took the lead in fighting the Taliban. In November 2001 the horse-riding Uzbeks joined with a U.S. Special Forces A-Team known as Tiger 02 and Air Force combat controllers in launching an offensive from their mountain enclave high in the Hindu Kush.[15] Using a combination of medieval-style cavalry charges and precision-guided joint direct attack munition (JDAM) air strikes, the Uzbeks and Americans broke out of the mountains and seized the holy shrine town of Mazar i Sharif from the Taliban on November 9, 2001.[16]

Dostum's bold seizure of Afghanistan's holiest spot struck panic into the heart of the Taliban. Their whole house of cards began to collapse from defections. By November 12 Kabul had fallen to the Northern Alliance, and the northern half of the country was soon thereafter liberated from the Pashtun-Taliban southerners. In Kabul crowds came out to kill stranded

Arabs, shave off their own beards, and tentatively begin to discard their burqas.

At this time Osama bin Laden and several hundred of his Arab followers made their way eastward to a base he had built in the Spin Ghar Mountains, which run along the Afghan-Pakistani border south of the Afghan border city of Jalalabad. There, at a place called Tora Bora (Black Dust), he planned to make a heroic stand. This was America's chance to send in U.S. special forces, including the 101st Airborne, Army Rangers, Delta Forces, and the Marines' Special Operations Command, to kill or capture the man who was the raison d'etre for the invasion of Afghanistan. Unfortunately, U.S. Central Command head Gen. Tommy Franks decided to rely on bombs and newly deputized Pashai and Pashtun tribesmen to flush out bin Laden. The Afghan tribesmen were taking money from both sides. They were subsequently bribed by bin Laden to allow him and his weary men to flee across the mountains to the nearby border of Pakistan.[17]

While America's tribal mercenaries turned a blind eye, bin Laden and his followers fled over the mountains into the Tirah Valley in the Pakistani tribal agency of Kurram. They were said to have been guided by the torches of sympathetic pro-Taliban tribesmen.[18] There they took advantage of their deep connections among the local FATA tribesmen, which went back to the anti-Soviet jihad of the 1980s, and the Pashtun tradition of *melmastiia* (obligation to host and protect a guest) to request sanctuary. Bin Laden and hundreds of his followers had succeeded in escaping to one of the most inaccessible places in the world among the fierce Orakzai and Afridi Pashtun tribes of Kurram Agency. Bin Laden's escape into the FATA was America's greatest blunder in the war on terrorism for this was sovereign Pakistani territory. President Musharraf could never allow U.S. troops to directly invade his nation in pursuit of their Muslim enemies for fear of backlash among his own people, who distrusted the Americans.

But this was not U.S. Central Command's only mistake. The hunt for the Taliban leader Mullah Omar was not going well either. An undisclosed host country (either Uzbekistan or Pakistan) had given the CIA permission to fly armed Predator drones from its territory on October 7, 2001, and they began flying on that very day.[19] The drones were actively hunting Omar and on one occasion spotted a Taliban convoy fleeing Kabul. A drone's high-resolution camera focused on the license plate of one car and found that it belonged to Mullah Omar. But instead of firing a missile at the car, the CIA controller asked for authority to fire on such a high-value target. The request eventually made its way from the duty officer at Central Command headquarters up to General Franks. Franks took the advice of his Judge Advocate General (JAG) Corps lawyer and decided not to fire on Omar.[20]

Having survived a close brush with death without even knowing about it, Mullah Omar subsequently escaped across the border into the Pashtun tribal zones of Pakistan. There he continued to rule the exiled Taliban as the "Commander of the Faithful" from Quetta, the sprawling capital of Baluchistan Province. The Americans' mistake proved to be catastrophic, for the messianic Omar was able to reunite and inspire his forces for years to come. Nearly ten years after his escape, one Taliban foot soldier explained his total awe of Mullah Omar, saying, "His words have a very powerful effect on us. We obey his orders, every Talib does, and we believe in him."[21]

Omar was not the only one who escaped a Predator owing to a reluctance to give firing orders. On as many as ten other occasions, high-value targets (HVTs) escaped after being spotted by drones, whose pilots had to wait for permission and further verification before firing. One officer captured his frustration over this sort of reluctance to fire when he said, "It's kind of ridiculous when you get a live feed from a Predator and the intel guys say, 'We need independent verification.'" Another Air Force officer told the *Washington Post*, "We knew we had some of the big boys. The process is so slow that by the time we got the clearances, and everybody had put in their 2 cents, we called it off."[22]

Clearly during the initial stages of the drones' utilization, the CIA was reluctant to use this latest killing tool to assassinate targets on the ground, even during a war. This infuriated Secretary of Defense Donald Rumsfeld. Rumsfeld was said to have kicked in a door when he heard how Mullah Omar got away. The secretary quickly changed the rules of engagement, making it easier for drone operators to take out their targets without having to go through an extended chain of command.[23] (The drones at this stage were being flown by the Air Force, which worked closely with the CIA to create targeting lists in a uniquely hybrid campaign.)

Rumsfeld may have done so just in time, for a drone subsequently spotted a large group of Arabs gathering at a three-story building south of Kabul known as the Yarmouk guest house. On this occasion not only did the drone operator receive instant permission from Langley and the Pentagon to fire on the target, but F-18 Hornets were called in to back up the Predator's Hellfire ammunition with bombs. The building was subsequently bombed, and the Predator fired missiles on trucks filled with panicked survivors trying to escape the scene of the devastation.[24]

The American drone pilots then eagerly listened in on the Arabs' radio chatter to see who had died in the attack. The CIA and National Security Agency (NSA) monitors were surprised to hear the Arabs bewailing the loss of someone described as "al Kumandan" (the Commander). There was only one al Qaeda leader who went by that name: the commander of al Qaeda's military wing, Abu Hafs al Masri (aka Muhammad Atef), the third highest ranking member of al Qaeda. Masri was not only head of al Qaeda's military operations, but he was also intimately involved in the 1998 embassy bombings in Africa that had killed hundreds of people, predominantly Africans. His daughter had recently married bin Laden's son to cement his relationship with his friend. The loss of the number three in al Qaeda was to be a tremendous blow to the organization and the first high-profile kill attributed to a Predator drone.

This victory appeared to give the CIA confidence, and as many as forty Hellfire missiles were fired by the agency's drones by mid-November 2001. Although the CIA worked closely with the Pentagon during the campaign, on several occasions Air Force officials monitoring Afghanistan noticed explosions and belatedly came to realize that they were caused by CIA drones hitting various targets in the country without informing them.[25] In this regard it should also be stated that the CIA was not the only organization deploying drones over Afghanistan during Operation Enduring Freedom. The U.S. Air Force had its own Predators, which were initially piloted remotely from Creech and Nellis Air Bases outside of Las Vegas, Nevada. (Air Force drones would later be flown out of Fargo, Holloman, March, Hector, Davis-Monthan, Beale, Ellington, Ellsworth, Fort Drum, Whiteman, Cannon, Eglin, Cannon and Hancock Airfields as well.) For their part, the CIA drones were controlled from Langley Air Base, 150 miles south of Washington, DC, and sent their remote images to the Global Response Center, on the sixth floor of CIA Headquarters in Langley.[26] The CIA drone campaign was run by the CIA Counterterrorism Center's Pakistan-Afghanistan Department.

Although the CIA and the Pentagon had clashed before 9/11 over who would pay for and utilize the drones, they appeared to have created synergy during Operation Enduring Freedom. Never was this better demonstrated than in March 2002's Operation Anaconda. This operation followed the capture of the southern capital of the Taliban, Kandahar, in December 2001. With the capture of Kandahar by opposition Pashtun tribal leader Hamid Karzai, the Taliban regime collapsed. Unable to control even the Pashtun tribal lands, most "village Taliban" returned to their homes while the hardcore Taliban fled across the border into the Pashtun tribal lands in Pakistan's FATA. There they received sanctuary in Bajaur Agency and North and South Waziristan in particular.

But a group of several hundred Arab al Qaeda fighters and Taliban members decided to resist the Americans from a remote Afghan mountain enclave near the Pakistani border at a place called Shah i Kot. In March 2002 the United States launched Operation Anaconda, an airborne assault to try flushing out the enemy in this area. The Americans who landed in Chinook helicopters were ambushed soon after arriving in the mountains and began to take casualties. One U.S. helicopter was downed, and its crew fought to survive the withering fire from the ambushers who were shooting at them from a nearby bunker. In desperation the U.S. troops called for air support to suppress enemy machine-gun fire coming from a ridge known as Takur Ghar. U.S. F-15s and F-16s were sent to save the team, and they attempted to take out the enemy bunker using cannon fire and bombs. But the supersonic jets that were screeching overhead at almost six hundred miles per hour could not be "walked to the target" via radio by the encircled team on the ground. The F-15s almost bombed the entrapped Americans by accident, and so their runs were called off for fear of killing their own men with clumsy five-hundred-pound bombs.

At this time a CIA Predator drone was hurriedly sent to the scene. The drone's cameras captured the scene below and relayed it back to CIA Headquarters in Langley. Using real-time high-resolution optics, the CIA operators were able to make out the Taliban bunker that had the American soldiers pinned down and to fire Hellfire missiles at it. When the bomb smoke cleared, a jubilant voice came from the trapped Americans on the ground: the enemy bunker had been destroyed.[27]

By now both the Pentagon and Langley had come to see the value of the Predator drone as both a battlefield weapon and a tool for targeted assassination. The combat career of the remote-controlled UAV had begun. But at roughly this time a CIA drone may have also made its first confirmed kill of an innocent victim, showing that the drones could be double-edged swords. Just prior to Operation Anaconda, a CIA drone had spotted three men walking in the hills of Zawhar Kili. Zawhar Kili, which is in the Afghan east, had been used in the 1980s as a base by the mujahideen, including bin Laden at one time. It had been heavily bombed by the United States during Operation Enduring Freedom.

In February the CIA drone operators noticed that the three men in the hills near Zawhar Kili appeared to be led by a tall man. The decision was made to fire a missile on the men gathered on the hill in the hopes that the tall man was bin Laden. The missiles were fired, and three men were killed instantaneously without ever knowing what happened to them. It was later announced that the CIA had fired on a figure suspected to be bin Laden himself. But when a *Washington Post* war correspondent rushed to the remote scene of the attack to investigate, he found that the man on the hill was not bin Laden at all. In fact he was a local villager named Mir Ahmad. He and his friends were scavenging for scrap metal from U.S. ordnance on the Afghan-Pakistani border when the drone found and killed them. The reporter described his findings as follows:

> "I was going past there toward Khost, and I heard the sound of an explosion," [a local villager] said. "The three were cut in half. They were just poor people trying to get money to feed their families."
>
> Khan said Ahmad had two wives and five children. The Pentagon has said that an unmanned Predator drone spotted a group of men at Zhawar, and that others seemed to be acting in a deferential manner toward one tall man. U.S. officials have said they received other, unspecified information that the men were al Qaeda leaders before giving approval to fire the missile.[28]

Another report from the area provides a heartbreaking account of how the local villagers dealt with the tragedy that had been inflicted on them by one CIA drone:

> They were there making a living, Gir's uncle said. His nephew "came down with a load of firewood from the mountains, and then said he was going out to collect some metal," Janat Khan said. "He said he'd be back soon."
>
> Late that afternoon, they heard the news of the missile attack, Janat Khan said. The men of the village gathered coffins and went to retrieve the bodies.
>
> "We were scared we would be bombed, but we had an obligation to bury them," Qosmat Khan said. They had to collect the pieces of two of the men. Daraz's body was intact, and he might have lived for a while, but he was dead when the village men arrived, said his brother.[29]

When subsequently asked about the unfortunate incident by a reporter who wondered if the errant strike presaged "some kind of public relations disaster," a defensive Donald Rumsfeld said, "I'm always concerned when there is an allegation made that suggests that some innocent person was— that an attack was inappropriate or that some innocent person was killed or injured. Obviously, anyone would be concerned about that."[30] Clearly everyone involved, from the secretary of defense to the head of the CIA, understood from this incident that the drones, for all their advanced optics and loitering capacity, were only as good as their intelligence. In their eagerness to kill bin Laden, the CIA drone operators had just killed three civilians in precisely the sort of mistake CIA director George Tenet had fretted about.

Thus, at a relatively early stage of the game, the CIA came to see the drones as an advantage and a liability. They were an unprecedentedly accurate tool for killing the likes of al Qaeda number three Muhammad Atef "al Kumandan," but they were still reliant on solid humint in order to be effective. The local Afghan governor at Zawhar Kili captured this dichotomy when he said, "We are happy that they [the Americans] came, and we are ready to help them. But the people are starting to get angry at them."[31]

As the ongoing war against al Qaeda and its Taliban allies gradually shifted across the border from eastern Afghanistan to the Pashtun tribal zones in neighboring Pakistan, the Americans would continue to wrestle with a paradox. While the war against the Taliban was transformed into a hunt for HVTs, it became obvious that America's most advanced weapon in the hunt for elusive terrorists might also be their worst enemy in the underlying battle to win the hearts and minds of the people of this volatile region.

5

Manhunt

It's a new kind of war. We're fighting on a lot of different fronts.
—U.S. secretary of state Condoleezza Rice

We cannot succeed in Afghanistan without shutting down those safe havens.
—Adm. Mike Mullen, the chair of the Joint Chiefs of Staff referring
to the Taliban and al Qaeda's sanctuary in the FATA

B y the spring of 2002 the Pentagon believed that the majority of hard-core Taliban members had been driven from Afghanistan. Those few Taliban members who were still sniping at U.S. troops in Afghanistan were described by Secretary of Defense Donald Rumsfeld as "dead enders."[1] Operation Enduring Freedom had been one of the most effective invasions in the history of the Afghan "Graveyard of Empires." It was now time to fill the void left by the collapse of the highly unpopular Taliban regime and rebuild the war-torn country that the United States had previously abandoned after the Soviets withdrew in 1989. The vast majority of Afghans wanted a return to school for their sons and daughters, new roads, demined fields, democracy, jobs, security, and an end to the Taliban's harsh misrule. To prevent the Taliban from coming back, the Americans and their NATO allies would have to rebuild the devastated nation from the ground up and offer this long-suffering people hope.

Unfortunately, the United States did not initially invest in Afghanistan's security and future because the Bush White House was adamantly opposed on principle to the notion of nation building. Candidate George Bush famously summed up his views on this topic when he stated, "If we don't stop extending our troops all around the world in nation-building missions, then we're going to have a serious problem coming down the road. And I'm going to prevent that."[2] For this reason the United States limited its military presence in the Texas-sized country of Afghanistan to less than ten thousand troops for the first few years of the conflict. The lack of a U.S. ground force allowed the down-but-not-out Taliban to begin to regroup in the Pashtun south as they awaited orders from their Pakistan-based leadership in the FATA and Baluchistan.

It also allowed the soon-to-be insurgent leader Gulbuddin Hekmatyar to return to the country. Hekmatyar, as mentioned previously, was the notorious fundamentalist mujahideen leader who had made a name for himself by throwing acid into the faces of unscarved women in Kabul, shelling the Afghan capital in the 1990s, and turning against his American sponsors during the 1991 Gulf War. In 1997 this Pashtun fundamentalist had been driven into exile in Iran by the Taliban, but in February 2002 he secretly returned to his old jihad stomping grounds in eastern Afghanistan. There he and his followers declared jihad on the American "infidels."

In May 2002 the CIA tracked Hekmatyar to a base in the forested mountains of Kunar Province on Afghanistan's northeastern border with Pakistan. There a CIA drone launched its Hellfire missiles at him and his followers. According to news reports, the missiles killed several of Hekmatyar's followers but failed to take out the leader himself.[3] From this point forward Hekmatyar is said to have changed "his location every five minutes" to avoid the drones.[4] The United States suspected that one of the locations Hekmatyar hid in was the FATA, a region he had used as a rear-area staging ground during the 1980s jihad against the Soviets.

Hekmatyar was not, however, the only ex-mujahideen warlord to ally himself with the Taliban as they licked their wounds in 2002 and prepared to launch an insurgency in the Pashtun belt of southeastern Afghanistan. The former CIA-backed mujahideen leader Jalaludin Haqqani had been courted by the CIA in 2001, but Haqqani saw the world in Manichean black and white jihad terms. To him the Americans were infidel invaders who needed to be expelled from Afghanistan just as the Soviets had been before them. As the American bombs fell in Operation Enduring Freedom, Haqqani and his followers fled from their base in the eastern Afghan province of Khost over the border into the neighboring Pashtun FATA tribal agency of North Waziristan. Haqqani had been based in this area during the 1980s, and he knew it well. There the so-called Haqqani Network regrouped in 2002 and awaited orders from its nominal leader, Mullah Omar. There were thus three Pashtun terrorist-insurgent networks, not including al Qaeda, based in the FATA and preparing to launch an insurgency against the Americans, who had too few troops to control this vast area of Afghanistan. Fortunately for the Taliban and al Qaeda, America's attention was about to be directed elsewhere—and America was about to snatch defeat from the jaws of victory.

As the Taliban and its allies regrouped in the FATA, the United States' attention began to be diverted toward the Iraqi ruler, Saddam Hussein. As early as February 2002 the United States began preparations for a massive invasion of Socialist-Baathist Iraq, which was said to have dangerous weapons of mass destruction. The head of U.S. Central Command (CENTCOM), Gen. Tommy Franks, summed up this redirection of NSA satellites, special operations troops, and Predator drones from Afghanistan to Iraq as follows: "We have stopped fighting the war on terror in Afghanistan. We are moving military and intelligence personnel and resources out of Afghanistan to get ready for a future war in Iraq." Franks described the downgraded mission in Afghanistan, a conflict that would soon become known as the "forgotten war," as a "manhunt."[5] Eventually 75 percent of U.S. drones would be transferred from Afghanistan to the new theater of action in Iraq.[6]

Shortly after the Iraq War started, one U.S. official said, "If we were not in Iraq, we would have double or triple the number of Predators across Afghanistan, looking for Taliban and peering into the tribal areas. We were simply in a world of limited resources, and those resources are in Iraq. Anyone who tells you differently is blowing smoke."[7]

In a fascinating incident in 2002, one of the newly transferred Predators entered an impromptu dogfight with an Iraqi MiG 25 Foxbat fighter plane. This was the first incident in which a UAV waged combat against a manned aircraft. Unfortunately for the drone, the faster-flying MiG jet fighter easily shot down the slow-moving Predator. The last thing the drone's remote pilots saw on their screens back in the United States was a missile shooting toward their aircraft as their own missile streaked toward the oncoming MiG. Then the screen went blank as the MiG's missile hit and destroyed the Predator.[8]

But the Predator's major headlines in 2002 came not from the new war in Iraq but from Yemen, where one of America's chief al Qaeda enemies, Qaed Senan al Harethi, known as the "Godfather of Terror," was planning terrorist attacks on the United States.[9] Harethi had been a wanted man ever since 2000, when he had helped plot the al Qaeda bombing of the USS *Cole*, which killed seventeen sailors. The U.S. government had requested that the Yemeni government arrest Harethi, and so the Yemenis sent their troops to capture Harethi, who was living in the Marib region of Yemen with allied tribes. But the arrest attempt ended in disaster as Harethi's tribal allies fought back and killed eighteen Yemeni police.[10] This incident vividly demonstrated the limitations of trying to arrest terrorists in remote tribal regions where the writ of the government is not recognized. Similar calls for the CIA to arrest suspected al Qaeda and Taliban terrorists in the remote tribal regions of Pakistan overlook the impossibility of penetrating these regions and arresting militants among thousands of their armed supporters.

Having failed to arrest Harethi, the CIA tried tracking him down using a Predator drone based in the French-garrisoned country of Djibouti as part of Combined Joint Task Force–Horn of Africa. The U.S. ambassador in Yemen also paid off some local tribesmen to keep track of Harethi's movements.[11] This two-tracked policy paid dividends on November 4, 2002, when the NSA's Yemen-based Cryptological Support Group tracked Harethi down by monitoring his cell phone. The NSA told the CIA, "Here's the general location. . . . We have the satellite phone, now go get the guy."[12]

A patrolling drone was rushed to the scene and spotted Harethi's vehicle. As Harethi and his compatriots made their way out into the desert, their SUV was suddenly hit with a powerful Hellfire missile that carried enough explosives to destroy a tank. Everyone inside was instantly incinerated. According to one report, all that remained of the travelers was "a mass of carbonized body parts."[13] Harethi's body was identified by a distinctive mark found on his dismembered leg.

But that was not all. It was later discovered that the CIA had killed two birds with one stone, for one of Harethi's companions was found to be none other than Kemal Darwish, the head of an al Qaeda sleeper cell based in Lackawana, New York. It so happened that Darwish was a naturalized U.S. citizen, which meant that the CIA had executed an American without providing him with a trial. U.S. government officials later claimed that they had not known in advance that Darwish was in the SUV, but as one official dryly put it, "it would not have made a difference. If you're a terrorist, you're a terrorist."[14]

The U.S. government's views on the assassination had clearly come a long way since George Tenet's cautious approach on the eve of 9/11, and this did not go unnoticed. The killing of an American citizen by the CIA raised eyebrows, as did the killing of a Yemeni citizen on Yemeni soil by a foreign intelligence agency. In many ways the Yemeni reaction to the targeted killing of two terrorists and their compatriots provided a foretaste of the reaction to the Americans' more systematic campaign of drone assassinations in Pakistan's remote tribal zones. Although the CIA's operations to kill Harethi had had the tacit approval of the Yemeni government, the actual drone strike had been carried out unilaterally. Some members of the Yemeni government expressed unease with the way the CIA had acted on their soil and the opposition parties howled in protest about the violation of Yemen's sovereignty.

The *Economist* presciently predicted, "The relentless advance of technology means the use of pilotless aircraft to hunt down terrorists will become more appealing to America. . . . The attack in Yemen will not be their last."[15]

Criticisms and fears of an expanded campaign did not, however, faze the Bush administration. President Bush subsequently declared, "The only way to treat them is [as] what they are: international killers. . . . And the only way to find them is to be patient and steadfast and hunt them down. And the United States of America is doing just that. We're in it for the long haul."[16] By defending the Yemen strike, the White House was serving notice of America's bold intention to use the latest technology available to track and kill al Qaeda members wherever they were hiding. One did not have to be versed in geopolitics to sense that that would ultimately take the Americans to Pakistan.

By this time it was clear to all that the main sanctuary that the United States was interested in denying to the terrorists was found in Pakistan's FATA region. Even as America began its massive invasion of Iraq, which would see U.S. CENTCOM deploy more than 150,000 troops (compared to a mere ten thousand U.S. troops in the larger country of Afghanistan), the CIA persistently kept up the manhunt for the original target of the war on terror, bin Laden.

When bin Laden and his number two, Ayman al Zawahiri, had themselves filmed in 2002 walking together in the mountains of what was presumed to be the FATA, the Americans were reminded of the importance of their unfulfilled mission. The U.S. State Department began to pressure the Pakistanis to do something about the fact that the world's most wanted terrorist and many of his accomplices were hiding out in territory nominally under their control. The White House wanted the Pakistani army to enter the autonomous tribal agencies and hunt down al Qaeda.

In July 2002 Pakistan responded to U.S. pressure by sending Pakistani troops to the FATA's Tirah Valley (Kurram Agency) for the first time ever. Their aim was to capture bin Laden, Zawahiri, and other foreign fighters and terrorists said to be hiding there. From Tirah Valley they proceeded into the Shawal Valley of North Waziristan and then into South Waziristan. This incursion into the FATA was made possible after long negotiations with prickly local Pashtun tribes that had always enjoyed their autonomy.

But as the Pakistani troops clumsily shelled compounds and *hujras* where the foreign terrorists and fighters were holed up, militants among the local Waziri and Mehsud Pashtun tribes rose up against them. The Pakistani army, it seemed, had inadvertently stirred the hornet's nest. By this time the press began to speak for the first time of the "Pakistani Taliban." These were ad hoc militias made up of local Pashtun fundamentalist militants, many of whom had gone to fight in Afghanistan on behalf of Mullah Omar's Taliban regime. They were joined by hundreds of Punjabi militants who had fled to Pakistan's remote agencies in 2002, when President Musharraf, at the behest of the Americans, banned the Pakistani jihadi groups that had fought in Kashmir.

In the FATA the militants began to organize resistance to the Pakistani army and to exert their own control. They extended their power through killings and intimidation. Any *malik* suspected of being a moderate, a secularist, or a supporter of the "apostate" Pakistani government was brutally murdered. Hundreds of tribal leaders were hunted down and killed in this assassination campaign, especially in the agencies of Bajaur and North and South Waziristan.[17] The Pakistani ISI's Taliban chickens had finally come home to roost. A typical report from the Pakistani press described this sort of creeping conquest by the informal Pakistani Taliban as follows:

"Talibanisation has taken strong roots in Orakzai and the region is now run by the Taliban council, which has introduced sharia (Islamic) law," tribesmen who have moved from Orakzai to escape Taliban-style rule told Daily Times on Tuesday. . . .

They said the Taliban council had banned women from travelling outside their homes without the escort of male family members. "There is a ban on music and dancing during wedding ceremonies; working of NGOs; and development works," they added. Each area now has its own Taliban chief and is patrolled by Taliban militants to keep the local population under the control of the TTP, the residents said.[18]

Another report by the British Broadcasting Company (BBC) captures the horrors faced by average tribesmen living under the harsh rule of the ever-expanding Taliban: "They [the Pakistani Taliban] were beheading people, they were shooting innocent people without any warning, they were terrifying us. . . . They were stopping our kids from going to school, they were kidnapping young boys. . . . With my own hands I have buried 18 people who were beheaded, even children. . . . They are not friends, they are not our allies, they're our enemies, they are criminals, they are gangsters."[19]

The Agence France-Presse similarly reported, "'Everything has changed in 10 years: most of the tribal leaders have been killed and the tribal system destroyed by the Islamists. We can't dance any more, or play music at weddings,' said Miramshah shopkeeper Qader Gul, 56. 'Anyone who protests risks having a member of his family kidnapped, beaten or killed.'"[20]

Having brutally achieved widespread control over many of the people and much of the countryside of the FATA by late 2003, the Pakistani Taliban prepared to resist the Pakistani army. The conflict came to a head in March 2004, when Pakistani infantry and mountaineering divisions tried to storm positions held by hundreds of Pakistani Taliban troops protecting al Qaeda fighters near the South Waziristan town of Wana. The Pakistani forces were convinced that al Qaeda leader Zawahiri was among the Taliban militants and were determined to capture or kill him. But the estimated three hundred to four hundred militants fought back ferociously from their compounds and killed between sixty and a hundred Pakistani troops, once again highlighting the difficulties in using conventional means to arrest terrorists in this sort of dangerous tribal region.[21]

To defuse the situation, the Pakistani government did an about-face and signed the first of many placating peace deals with the emboldened Pakistani Taliban in April 2004. The man who emerged to sign the so-called Shakai Peace Accords on behalf of the local Taliban was a charismatic South Waziristani Pashtun tribesman named Nek Muhammad. He was said never to have retreated in battle and had earned the nickname "Bogoday" (the Stubborn One).

Nek Muhammad had been forced to flee Afghanistan in 2001, when the United States invaded and the Taliban army melted away under U.S. bombs and Northern Alliance Uzbek cavalry charges. At this time Muhammad led many of his Arab and Uzbekistani jihadi compatriots across the Pakistani border to his native South Waziristan. There he and his fellow Waziri Pashtun tribemen offered the foreigners the hospitality of *melmastiia*. The Pakistani government insisted that the foreigners be turned over to them, but Muhammad adamantly refused to surrender fellow Muslim fighters to the U.S. puppet government of President Musharraf.

As previously mentioned, soon thereafter the Pakistani government signed the Shakai Peace Treaty with Nek Muhammad, who interpreted this treaty as a capitulation by the Pakistani army. To sign the treaty, a Pakistani general actually flew to Nek Muhammad's territory and symbolically recognized his authority by placing a garland of flowers around his neck. As part of the treaty, Muhammad was told he had to turn over foreigners in his territory, but he promptly declared that there were no foreigners in his lands to turn over.[22] Nek Muhammad had thus broken a major stipulation of the treaty before the ink on it had dried. As if to pour salt in the Pakistani army's wounds, at this time he also promised to launch terrorist attacks throughout Pakistan to punish the government for invading South Waziristan in the first place.[23]

Like Mullah Omar, who refused to turn over bin Laden back in 2001, Nek Muhammad had made a decision to stand up to the Americans. No amount of military incursions or appeasing peace treaties could convince the Stubborn One to turn over his Arab and Uzbekistani guests. Instead, he began launching small-scale attacks on Pakistani troops and declared that his people had "fought a jihad against the Russians and before them the British. Now that the Americans are here we will wage jihad against them."[24]

As the Pakistani Taliban metastasized and began to threaten both Pakistan and the United States, it became clear that Nek Muhammad was the major impediment to joint Pakistani-U.S. operations designed to flush out al Qaeda operatives hiding in the tribal areas. For his part Muhammad seemed to relish the notoriety that came from his decision to stand up to both the Pakistani army and the Americans. In many interviews by the media, he boldly promised to wage total jihad against the Pakistani government and its U.S. allies. Safe in his inaccessible tribal land, surrounded by thousands of armed followers, Muhammad seemed untouchable.

Then, on June 19, 2004, the Pakistani English-language newspaper *Dawn* reported that the Stubborn One had been killed two days earlier:

Security forces have killed Nek Mohammad and four other tribal militants in a missile attack on a village in Wana, the regional headquarters of the South Waziristan Agency. "Nek Mohammad was suspected to be present in a hideout with his associates and our security forces acted swiftly on the information and that is how he was killed," a [Pakistani] military spokesman said on Friday.

Residents of Shah Nawaz Kot, a small hamlet about two kilometers south of Scouts Camp, said the 27-year-old militant was killed when his hideout was hit by a missile. The attack was reportedly carried out at 10pm on Thursday night when Nek Mohammad was taking dinner along with his colleagues in the courtyard of the house of his long-time friend, the late Sher Zaman Ashrafkhel, an Afghan refugee from his Ahmadzai Wazir tribe.

Witnesses said that Nek Mohammad's face bore burn marks and his left hand and leg appeared to have been badly injured in the explosion. "Why aren't you putting a bandage on my arm," were his last words, those accompanying Nek Mohammad to the hospital quoted him as saying.[25]

Pakistani security forces were thus said to have "acted swiftly" to kill Muhammad with a surprisingly accurate missile strike at night. But the Pakistanis had never carried out such a precise, nighttime strike before, and military experts doubted they had the technology to do so. Pakistani reporters began asking questions in hopes of clarifying how exactly the Pakistani army had carried out this feat. The *Dawn* story suggested that although the Pakistanis claimed the strike as their own, in actuality it was conducted by an American drone flying from Afghanistan. After dutifully sharing the Pakistani military's official version of the death of Nek Muhammad, *Dawn* contradicted the official claim:

> Witnesses said that a spy drone was seen flying overhead minutes before the missile attack. There were also reports that Nek Mohammad was speaking on a satellite phone when the missile struck, fuelling speculations that he might have been hit by a guided missile. The precision with which the missile landed right in the middle of the courtyard where Nek Mohammad and his colleagues were sitting, lent credence to the theory. Locals said that the missile created a six feet crater.
>
> An associate of Nek Mohammad, who called the BBC Pushto office in Peshawar, also said that the tribal militant had been killed while speaking on a satellite phone. The Inter-Services Public Relations (ISPR) declined to speculate on how the militant had been killed.
>
> "We have various means and a full array of weapons at our disposal. We have artillery that can fire with precision and we have helicopters with night vision capability which can fire guided missiles. But I am not going to give out operational secrets on how he was killed," ISPR Director-General Maj-Gen Shaukat Sultan told Dawn by phone from Islamabad.
>
> "Absolutely absurd," was his response when asked about rumors that Nek Mohammad had been killed with the US assistance. "Intelligence is like a jigsaw puzzle, it does not come from a single source on a single time," Gen Sultan said.[26]

Despite the Pakistani military's claims that it was "absolutely absurd" that the strike had been carried out with U.S. assistance, the eyewitness quoted in the *Dawn* article claimed to have heard a drone flying overhead just prior to Muhammad's assassination. The witness also claimed that the victim was speaking on a satellite phone at the time of his death. It should be recalled that Harethi was previously tracked down in Yemen by the NSA after he used a cell phone. It seems likely that Muhammad was killed by a drone after making the same mistake. Clearly the Pakistani military was trying to deflect criticism away from itself for collaborating with the infamous CIA by claiming the strike as its own.

Regardless of who actually assassinated Muhammad, no anti-American protests followed his death because the Pakistanis were successful in their efforts to take credit for the assassination. But CIA drones were operating from Pakistani soil at this time; in January 2003 a Predator had crashed after takeoff at the Jacobabad airfield in Baluchistan.[27]

The drones' presence was even more evident almost a year after Muhammad's death, when another terrorist was killed in the region, this time a top al Qaeda agent named Haithem al Yemeni. Yemeni was next in line to replace the number-three man in al Qaeda, Abu Faraj al Libbi, who had been arrested by the Pakistanis. Upon his arrest, Libbi had been interrogated and was said to have given details of Yemeni's movements. These were then given to the CIA, which used a drone to assassinate Yemeni while he was riding in a car driven by a local Taliban warlord named Samiullah Khan.[28] Yemeni was said to have been killed near Mir Ali, North Waziristan, on or before May 13, 2005, after being under surveillance by the CIA for more than a week. CIA officers had monitored his movements for several days in the hopes that he would lead them to bin Laden or Zawahiri, but decided to kill him after fearing that they would eventually lose him.[29]

At the time, a Cable News Network (CNN) analyst pointed out that if the word of Yemeni's death at the hands of a CIA drone spread, it could "create political problems for the Pakistani government, which has been quietly cooperating with U.S. efforts to round up or kill al Qaeda operatives."[30] News of the assassination would of course be a public relations disaster for the Pakistani government, whose citizens were among the most anti-American people in the world, should it become known that government officials had given the CIA carte blanche to hunt its enemies on their territory

A National Broadcasting Company (NBC) source shed some important light on the relaxed procedures behind the covert campaign: "Sources told NBC News that the CIA has all the approvals necessary within its counterterror center in Langley, Va., to fire missiles within Pakistan when an al-Qaida target is spotted. The agency does not have to check with the White House or with Pakistani authorities or the CIA director. The purpose is to expedite rapid action in the field in case the opportunity is time sensitive."[31] This little noticed news was actually very important. In effect the Pakistanis and the White House had given the CIA's Counterterrorism Center (actually located in Tysons Corner, Virginia, not Langley) the green light to fire at will in the effort to assassinate targets in Pakistan, without seeking approval from either government.[32]

Years later President Musharraf admitted to giving the CIA permission to carry out surveillance with drones but denied giving the organization the right to fire missiles.[33] This retroactive denial does not, however, seem plausible in light of the subsequent drone campaign carried out during his tenure as president. It is more likely that Musharraf gave the Americans full authority to hunt terrorists as part of his new relationship with them but kept this agreement hidden from his own people. The Pakistani authorities had thus come a long way since their days of supporting the Taliban. The CIA had also come a long way since September 4, 2001, when Director Tenet worried about issues related to how the U.S. government and others would react when Arab terrorists began being assassinated. Clearly the CIA was now actively hunting terrorists operating from the autonomous "Talibanistan" zone in the FATA, which was officially claimed by the Pakistani government.

Not all voices were in favor of this new cooperation in killing terrorists in Pakistan. The human rights group Amnesty International issued a statement after the killing of Haithem al Yemeni that said,

> Amnesty International fears that, if the circumstances of these killings have been reported accurately, the USA has carried out an extrajudicial execution, in violation of international law. Amnesty International reminds the USA that it has condemned such unlawful actions when carried out by other states in the past. It calls upon the USA to end immediately all operations aimed at killing suspects instead of arrest-

ing them, investigate all past suspected cases of extrajudicial executions, and revoke all orders that may allow extrajudicial execution.[34]

Despite such calls for an end to the assassination campaign, the cooperation between Pakistan and the United States in carrying out the killings was expanded. In fact, NBC reported that the drone attack on Yemeni had actually been preceded by several strikes on sites in Pakistan described as terrorist training camps.[35] This was not surprising considering the growth in terrorist camps in the FATA since the overthrow of the Taliban regime in Afghanistan in 2001. According to one source, the Taliban and al Qaeda eventually established as many as 157 terrorist training camps throughout the FATA.[36] Numerous terrorist attacks, including the 2002 Bali bombings, the 2003 Istanbul bombings, and the 2005 bombings in London, could be traced to these camps. The CIA feared that future 9/11s were being plotted in the terrorist camps in the FATA and continued to monitor them by means of high-flying drones.

Word that the reconnaissance drones were armed may not, however, have gotten out yet. That would change with the next assassination: the killing of the new number three in al Qaeda, Abu Hamza al Rabia, in November 2005. Rabia had been involved in two terrorist plots to kill President Musharraf.[37] Before his death, one Pakistani source stated that he had been playing hide-and-seek with the CIA.[38] In fact, he had survived a drone strike on November 5, 2005, that had reportedly broken his leg and killed his wife and child.[39] Then he and four bodyguards, two of them Syrian, were mysteriously killed on November 30, 2005, at 1:45 a.m. in a village located near the capital of North Waziristan, Miranshah. Rabia died in an unexplained explosion that destroyed the mud-walled compound where he was living at the time.

In response to media inquiries about the possible role of a Predator in Rabia's death, a Pakistani intelligence official ambiguously stated that the terrorist had been killed while "working with explosives."[40] But such comments directly contradicted eyewitnesses who reported seeing white streaks flying from the sky into the house.[41] One witness reported, "I heard more explosions and went out to the courtyard, and when I looked up at the sky, I saw a white drone. I saw a flash of light come from the drone followed by explosions."[42]

The truth came out when a local Pashtun freelance journalist named Hayatullah Khan was hired by the U.S. television show *Frontline* and traveled to the scene with a camera. There he filmed local villagers holding pieces of shrapnel with the words "U.S. Guided Missile, Contact Serno, AGM-114" clearly visible on them.[43] Hayatullah's pictures were subsequently broadcast around the world and deeply embarrassed the Pakistani government, which had sought to downplay its covert working relationship with the CIA.

Afterward, Hayatullah feared that he had incurred the wrath of the Pakistani authorities and told reporters, "When I published the picture of the missile I knew the government would harm me." Hayatullah's premonition proved to be correct; he was subsequently kidnapped by unknown people. After his tribe launched a massive search for him that included questioning the local Taliban, Hayatullah's body was found with five bullet holes in it; his hands were bound by government-issued handcuffs.[44]

The Pakistani security forces were clearly sensitive to the issue of the drone strikes—so sensitive that they were willing to resort to murder to keep secret their role in them. Hundreds of Pakistani journalists subsequently protested the killing of one of their own by the country's security forces. The Pakistani government may have also tried to keep the lid on the news that the son and nephew of the man who had given hospitality to Rabia had also been killed in the drone strike. These two were possibly the second pair of collateral damages of the Pakistani drone campaign (the first being Abu Hamza al Rabia's wife and child, who were killed in the earlier strike).[45]

For their part the Americans were thrilled with the death of the third-highest ranked al Qaeda terrorist in Pakistan. National Security Adviser Stephen Hadley said, "This is law enforcement, this is not assassination. This is going against the leadership of an organization that has declared war on the United States."[46] By this time the Pakistanis had lost more than 250 soldiers fighting against the Taliban while searching for al Qaeda terrorists and were still willing to continue to live with the lie that the terrorists' deaths were carried out by "car bombs" or "explosive mishaps."[47] Many in the ISI and military who had previously supported the Taliban had been purged. President Musharraf clearly saw the Taliban and al Qaeda (which had tried to assassinate him on several occasions) as a threat. While there had been cases of Pakistani frontier soldiers firing on U.S. troops in support of newly reorganized Afghan Taliban insurgents along the border, the Pakistani military and intelligence appeared to be tentatively throwing in its lot with the Americans.

This newfound cooperation would be strained following the next drone strike, which was aimed at none other than Ayman al Zawahiri. As it transpired, the Pakistanis had gained considerable insight into the al Qaeda number two's movements following the interrogation of the arrested al Qaeda operative Abu Faraj al Libbi. According to Libbi, Zawahiri had found sanctuary among militants in the village of Azam Warsak in South Waziristan. He had in fact married a Pashtun woman from the region to cement his relationship with the local tribesmen. In mid-March 2004 the Pakistani army bombed Azam Warsak and later received word that Zawahiri had been injured in the bombing.[48]

Zawahiri then fled north to the tribal agency of Bajaur and found refuge in the village of Damadola, which was controlled by pro-Taliban militants led by Sufi Muhammad. But the CIA was not far behind, and on or around January 5, 2006, it fired a missile into a compound believed to be housing Zawahiri. The al Qaeda operative was not killed in the attack, but eight family members of a "militant cleric" who was said to have given him hospitality were.[49]

Then, just before January 13, 2006, word came that Zawahiri had been invited to a compound in Damadola where known terrorists had met before. On this occasion he was to celebrate the Muslim feast of Eid al Adha and have a "terrorism summit."[50] Libbi had told his interrogators that such meetings had happened in the locale before. There were "strong indications that was happening again," and "all the intelligence signatures" said that Zawahiri himself would be there for the meeting.[51] It also emerged that some of the men said to be in attendance would be important al Qaeda operatives who had been tracked by the ISI since they had entered Pakistan from Afghanistan.[52]

It was exactly the opportunity the CIA had been waiting for. The new CIA head, Porter Goss, personally gave the order to launch a barrage of drone missiles into the compound and was also said to have notified White House officials of the impending attack.[53] According to one report, four drones took part in the subsequent strike, probably the largest flight of Predators involved in an attack to that date. As many as ten missiles were fired into the buildings thought to be housing Zawahiri and his comrades, according to Pakistani sources (this would indicate that as many as five Predators were actually involved in the attack since each Predator can carry only two Hellfire missiles).[54] Once again CIA screens in Virginia lit up with explosions as the Predators sent back live feed from the FATA to their handlers.

Then reports that several high-ranking al Qaeda members had been killed began to come in. But even as the American government triumphantly proclaimed that it had killed several top terrorists, it became clear that something had gone horribly wrong.[55] Although the Pakistanis initially claimed that the Americans had missed Zawahiri by no more than an hour or two, it turned out that the cautious al Qaeda number two had not attended the meeting in the first place. Instead he had decided at the last minute not to visit the compound and had sent four of his representatives in his stead.[56] The representatives were there when the strike occurred. According to one Pakistani source, eleven "extremists," including seven foreigners, were killed in the attack.[57]

And they were not the only ones. One survivor, Shah Zaman, who lost two sons and a daughter in the attack, said, "I ran out and saw planes. I ran towards a nearby mountain. When we were running we heard three more explosions and I saw my home being hit."[58] After the smoke cleared, it became apparent that between thirteen and thirty people had been killed in the Hellfire missile barrage, among them Zawahiri's four representatives, said to be Egyptians.[59] Their bodies were taken away by al Qaeda militants before they could be identified. The rest of the victims appeared to have been civilians (although ethicists were to ponder how innocent a civilian was if he or she was giving sanctuary or meeting with known terrorists). Regardless, as many as fourteen of the civilians killed in the drone strike were said to have come from the same family, and they included five children and women.

As word got out that three separate compounds with civilians in them had been destroyed in Damadola, the Pakistani government (whose agents had of course tracked the four important al Qaeda targets to the houses) gave a rather guarded statement: "According to preliminary investigations there was foreign presence in the area and that, in all probability, was targeted from across the border in Afghanistan."[60] This weak explanation both continued the myth that the drones were "invading" Pakistani airspace (when they were in all probability based in Shamsi and Jacobabad in Baluchistan Province) and pointed out that the Americans had actually been trying to target foreign terrorists, not Pakistani civilians.

The explanation did nothing to mollify average Pakistanis, who were infuriated by the summary execution of their compatriots by a foreign intelligence service that was notorious in the country. Interestingly, on this occasion the Pakistani government did not bother trying to cover up the CIA strike or deny that it had happened. The cat was now out of the bag; the CIA was actively killing suspected terrorists and even innocent bystanders on Pakistani soil.

The resulting uproar was not entirely unexpected considering the Pakistani public's deep distrust of America. Some ten thousand people rallied in the port of Karachi, Pakistan's largest city. There they chanted "Death to America" and "Stop bombing against innocent people."[61] Hundreds also gathered in the capital of Islamabad, as well as in Lahore, Multan, and Peshawar, the capital of the Pashtun-dominated North-West Frontier Province. Many burned U.S. flags and demanded the withdrawal of U.S. troops from Afghanistan, while others torched the office of a U.S.-backed aid agency. In the FATA itself a group of approximately eight thousand tribesmen chanting "God is great" protested the strikes.[62] The CIA had caused a major public relations disaster with a key strategic ally, all for the sake of killing four unknown al Qaeda operatives.

Not one to miss an opportunity, the intended target of the strike, Zawahiri, quickly released a video response to the attempt on his life. Gloating about his survival, Zawahiri addressed President Bush directly: "Butcher of Washington, you are not only defeated and a liar, but also a failure. You are a curse on your own nation and you have brought and will bring them only catastrophes and tragedies. Bush, do you know where I am? I am among the Muslim masses."[63]

Zawahiri was not the only one reacting; the Pakistani government was terribly embarrassed by the incident. As Pakistani protestors demanded the resignation of President Musharraf, the Pakistani government strongly condemned the strike to the U.S. ambassador. Pakistani information minister Sheikh Rashid Ahmed called the air strike "highly condemnable" and said the Pakistani government wanted "to assure the people we will not allow such incidents to reoccur."[64]

The Damadola strike also appeared to have galvanized an antidrone sentiment among the antiwar movement in the United States as well. One American reverend wrote a blog post for *CounterPunch* titled "Remember Damadola":

> America's moral decline is seen in the widespread mainstream accept-ance of the murder of women and children and other persons by our government in our name. How little the lives of distant and different human beings seem to count reveals how much we are falling from grace. . . . President Bush expressed no sympathy to the families and friends of the dead Damadola villagers. No apology to their loved ones, nor to the people of Pakistan. No restitution offered. No explanation given to morally concerned American citizens about the killing of innocent human beings.[65]

In fact no apology was forthcoming from the U.S. government. On the contrary, one Pentagon official said, "The message to [the Pashtuns of the FATA] is, 'You have to take a new measure now: your families are not safe if you protect the terrorists.'"[66] And despite all the public clamor in Pakistan and the growing antidrone movement in the United States, President Musharraf remained committed to his newfound alliance with the Americans who had by this time paid his country $10 billion in aid. (This total would ultimately rise to more than $20 billion.)

An interesting report in *Time* magazine from the period provides insight into the relationship between the Americans and the Pakistanis after the Damadola strike:

> Although the missile strike provoked a round of protests in Pakistan's tribal areas that forced President Pervez Musharraf to distance his gov-ernment from the operation, cooperation between the U.S. and Pakistan in the hunt for bin Laden has quietly deepened. A Peshawar-based Pakistani intelligence official speaking on condition of anonymity says Washington has an understanding with Islamabad that allows the U.S.

to strike within Pakistan's border regions—providing the Americans have actionable intelligence and especially if the Pakistanis won't or can't take firm action. Pakistan's caveat is that it would formally protest such strikes to deflect domestic criticism.[67]

To support U.S. operations, the Pakistanis joined with the CIA to create a network of local spies and informants who were actively involved in hunting for al Qaeda HVTs. One Pakistani security officer told *Time* magazine that the CIA had installed sophisticated surveillance equipment in several ISI offices to monitor radio and Internet communications between al Qaeda members.[68]

The results of this continuing collaboration remained mixed, according to a 2006 *Los Angeles Times* report based on interviews with CIA operatives who served alongside the ISI: "American spy agencies depend heavily on cooperation from Pakistan's Inter-Services Intelligence agency, elements of which are believed to have long-standing ties to the Taliban. Underscoring the lack of trust, a former high-ranking CIA official said that the United States typically gives the Pakistani government less than an hour's notice before launching a Predator missile strike, largely out of fear that more time might allow ISI sympathizers to tip off targets."[69]

At roughly this time a drone made headlines in the war effort that was the Americans' main focus at this time, Operation Iraqi Freedom. In June an Air Force drone was used to monitor a safe house where Iraq's most notorious terrorist leader, Abu Musab Zarqawi, the "Butcher of Baghdad," was said to be hiding. The loitering drone sent back live video of the house and finally recorded Zarqawi entering the isolated farmhouse on June 7. On the basis of this video, the Air Force directed two F-16s to bomb the house with five-hundred-pound bombs since the Predator's two Hellfire missiles were not deemed powerful enough to destroy the house. Zarqawi, a bloody terrorist who had introduced suicide bombings, sectarian death squads, and beheadings to the insurgency in Iraq, was killed.[70] The death of this charismatic leader who had created al Qaeda in Iraq in 2004 caused a collapse of leadership in his organization. Although several lackluster leaders came after him, none of them had the leadership qualities and fame of the notorious Zarqawi.

Despite this success in Iraq, the situation on the Afghan-Pakistani front was deteriorating. By the spring of 2006 the Afghan Taliban had fully reinfiltrated Afghanistan from their safe havens in Pakistan's FATA and had launched a full-blown insurgency-terrorism campaign. At this time the most effective Taliban commander was none other than the former CIA-sponsored mujahideen leader based in North Waziristan, Jalaludin Haqqani. His insurgents introduced Iraqi-style terrorism, complete with unbearably gruesome, videotaped beheadings, suicide bombings, and improvised explosive device attacks, to Afghanistan. Safe from American troops in their North Waziristan sanctuary, Haqqani's terrorists began a concentrated campaign to wreck Afghan president Hamid Karzai's dreams of bringing peace and prosperity to his war-weary people.

The situation on the Pakistani side of the border was not improving either. Having lost hundreds of their troops to tough Pakistani Taliban and al Qaeda fighters, the Pakistani authorities decided to sign yet another face-saving peace treaty with their enemies in October 2006. The so-called Waziristan Accords in essence led to the creation of a Taliban state often called the Islamic Emirate of Waziristan. An article in the *Weekly Standard* vividly described the accords, which recognized the de facto secession of Waziristan from Pakistan, as follows:

> Yet even in the wake of Pakistan's earlier surrender of South Waziristan, this new agreement, known as the Waziristan Accord, is surprising. It entails a virtually unconditional surrender of Waziristan.
>
> The agreement is, to put it mildly, a boon to the terrorists and a humiliation for the Pakistani government.
>
> Immediately after the Pakistani delegation left, al Qaeda's flag was run up the flagpole of abandoned military checkpoints, and the Taliban began looting leftover small arms. The Taliban also held a "parade" in the streets of Miranshah. Clearly, they view their "truce" with Pakistan as a victory. It is trumpeted as such on jihadist websites. . . .
>
> The ramifications of the loss of Waziristan are tremendous. The region that Pakistan has ceded to the Taliban and al Qaeda is about the size of New Jersey, with a population of around 800,000.[71]

Despite the billions of U.S. taxpayer dollars invested in Pakistan, the Pakistanis had surrendered control of their territory in Waziristan to the Taliban and al Qaeda. The terrorists were now free to plan more terrorist attacks in Afghanistan and against the West from their autonomous state located right on the Afghan border.

This development could not have come at a worse time and represented a victory for the Pakistani Taliban, all three Afghan terrorist factions, and of course al Qaeda. Al Qaeda's vitality in the FATA was vividly demonstrated in the fall of 2006, when British security agents foiled an attempted al Qaeda plot to use liquid explosives to blow up as many as ten passenger jet airliners. To this day airline passengers cannot store liquids in carry-on baggage as a result of this plot, which would have seen hundreds of people killed when their planes exploded in midair. The planners of the liquid bomb plot had received direct orders from the al Qaeda leadership in the FATA.[72] This close call and links to the FATA further galvanized the CIA's efforts to assassinate al Qaeda leaders before they could organize additional mass-casualty terrorism attacks in the West.

Then, in October 2006 the Pakistanis claimed to have located Zawahiri, once again in the Damadola region of the Bajaur Agency. He was said to be under the protection of two local pro-Taliban militants, named Maulvi Liaqat and Maulvi Faqir Mohammad, who operated a madrassa in the Chenagai suburb of Damadola. (*Maulvi* is a term for a high-ranking mullah.) This madrassa and the region in general were known as a hotbed for jihadists, who were said to cross the border to fight U.S. troops in the neighboring Afghan province of Kunar. Three thousand local militants had recently gathered at the madrassa to express their solidarity with Mullah Omar and Osama bin Laden, whom they claimed were their "heroes."[73]

As a result, the NSA began monitoring the region with high-resolution satellites and found evidence of militants from the madrassa training for combat.[74] The CIA also began to monitor the "terrorist training facility" with Predator drones, which locals saw flying in the area in late October (remember that drones can provide more than twenty-four hours of close-up surveillance in ways that satellites, which have to rely on orbits, cannot).[75] Finally, at around 5:00 a.m. on October 30, 2006, the decision was made to attack the madrassa with drones. A fusillade of Hellfire missiles was launched into the seminary. The aftermath was worse than it had been on any previous or subsequent strikes. As local villagers shifted through the rubble, they found as many as eighty-two people dead. Among them was one of the militant heads of the "terrorist compound," Maulvi Liaqat.

Local villagers, however, said that at least twelve of the victims who died in the attack were teenagers—which in and of itself would not exclude them from the ranks of the militants, but this information was nonetheless damning in the eyes of many.[76] A report by the Pakistani newspaper the *News*, titled "Most Bajaur Victims Were under 20," was even more damning and claimed that "one of the deceased was only seven-years old, three were eight, three nine, one was 10, four were 11, four were 12, eight were 13, six were 14, nine were 15, 19 were 16, 12 were 17, three were 18, three were 19 and only two were 21-years old."[77] If accurate, this claim would indicate that the CIA had targeted a school and primarily killed young students.

Pakistan's military spokesman, Maj. Gen. Shaukat Sultan, claimed that all those who died in the madrassa attack, regardless of their ages, were militants and explicitly stated that there was "no collateral damage."[78] Although some have tried to argue that the victims in the second Damadola strike were all innocent students, the fact that leaders from the mosque subsequently vowed to send out "squads" of suicide bombers to punish the Pakistani military for its role in the strikes would suggest otherwise.[79]

Regardless, the slain Maulvi Liaqat's chief ally in the village, Maulvi Faqir Mohammad, stirred up local anger at the Americans and the Pakistani government. During a speech to as many as ten thousand mourners, Mohammad declared, "The government attacked and killed our innocent people on orders from America. It is an open aggression."[80] He then promised to continue to wage jihad against the Americans and Pakistani government. Such threats became reality a week later when a suicide bomber dressed in a shawl rushed into a training area where Pakistani soldiers were doing their morning exercises and blew himself up. The result was devastating. Forty-two soldiers were killed and twenty wounded in the largest ever suicide bombing of Pakistani troops.

Given that hundreds of troops had already been lost in the fight in Waziristan, the loss of forty-two soldiers in the bombing prompted further debate throughout Pakistan. One Pakistani general argued, "We need a major rethink of the entire policy. We should not be fighting America's war. We have to solve our own problems. If we are dictated to by outsiders it will end up like Iraq or Afghanistan."[81]

Still, the drone strikes went on. In January 2007 the Pakistani military joined the hunt for al Qaeda, and there were reports that laser-guided bombs dropped by Pakistani jets had hit al Qaeda compounds in the tribal region of South Waziristan. This angered local tribal leaders, who felt that previous peace treaties with the Pakistani military had forbidden such activities.[82] Perhaps in response to pressure from the tribes, the Pakistani army in 2007 signed yet another peace treaty, this time with Maulvi Faqir Mohammad. The treaty was essentially a capitulation that ceded the northern tribal agency of Bajaur to Mohammad and his militants. This very same Faqir Mohammad had offered protection to Zawahiri, sent thousands of local tribesmen to support the Afghan Taliban in 2001, and openly declared his support for Osama bin Laden and Mullah Omar, whom he labeled "heroes of the Muslim world."[83]

By this time the Pakistani government had essentially ceded three tribal agencies to the Taliban: North Waziristan, South Waziristan, and Bajaur. This Pakistani surrender did not, however, bring a halt to CIA drone strikes in the region. On April 27, 2007, a madrassa in the village of Sadigi, North Waziristan, belonging to a pro-Taliban leader named Maulana Noor Mohammad, was hit with missiles that killed four.[84]

The next drone strike took place on June 20, 2007, in the village of Mami Rogha in North Waziristan and led to the death of at least twenty people in what was described as a terrorist training camp.[85] At roughly this time America's sixteen intelligence organizations had also produced their annual National Intelligence Estimate (NIE), which declared that al Qaeda had "protected or regenerated key elements of its Homeland attack capability, including: a safe haven in the Pakistan Federally Administered Tribal Areas (FATA), operational lieutenants, and its top leadership." This Pakistani safe haven "allowed al-Qaeda to act with virtual impunity to plan, train for, and mount attacks."[86] One senior military officer would describe the region as "the epicenter of terrorism in the world," and CIA director Michael Hayden later said, "They [al Qaeda] were coming at us. They were a threat to the homeland."[87]

But even as the CIA continued its efforts to convince the Pakistanis of the dire threat the al Qaeda–Taliban nexus posed, the troubles of Pakistan's remote frontier finally began to affect those Pakistanis who had turned a blind eye to the rise of the militants. In July 2007 militants from a major mosque, known as the Lal Masjid (Red Mosque), in the capital, Islamabad, began to pour into surrounding neighborhoods and clash with storeowners who they claimed were selling "pornographic" digital video discs (DVDs) and videotapes. They also kidnapped local women they described as "prostitutes" and seized control of a nearby government building. Many of the militants were women clad in hijabs and armed with automatic weapons. The militants, who were mainly conservative Pashtuns, then called for the enforcement of strict shariah law—the sort that was already being harshly enforced in Bajaur and Waziristan—in Pakistan. The heads of the Lal Masjid, known as the Ghazi brothers, also threatened to unleash suicide bombers on the capital if the government refused their demands to introduce shariah law nationally.

The Pakistani government had no choice but to act against this blatant challenge to its authority in the heart of the capital and sent troops to surround the mosque. President Musharraf proclaimed, "In the garb of Islamic teaching they have been training for terrorism. . . . They prepared the madrassa as a fortress for war and housed other terrorists in there. I will not allow any madrassa to be used for extremism."[88]

Pakistani troops then stormed the Lal Masjid and fought with the militants for several days before gaining control of the building. Ninety-one militants and eleven Pakistani soldiers were killed in the fighting. The outcome of the fighting originally seemed like a victory for the Pakistani government, but when word of the siege reached the Taliban in its autonomous "emirates" in Bajaur and Waziristan, it declared an end to the tentative "truce" with Islamabad and the beginning of jihad on the Pakistani state. The independent lands of Talibanistan were now officially at war with the Pakistani state, and Pakistanis could no longer pretend that the war on the terrorists was purely in the interest of the Americans.

Among the Taliban's first act was to send scores of suicide bombers against civilian and military targets, murdering more than a hundred in less than a week.[89] The Pakistani military was thus forced to respond and invaded

Waziristan, setting off battles that led to the deaths of hundreds of Taliban fighters, Pakistani troops, and civilians. The Taliban responded to this assault by invading the so-called settled lands of the Pashtun-dominated North-West Frontier Province and seizing control of Swat Valley, just a hundred miles to the west of Islamabad.

As this "creeping Talibanization" was being carried out by the Pakistani Taliban, the CIA launched a drone strike on the most effective of all the Afghan Taliban militants hiding out in North Waziristan, Jalaludin Haqqani. The November 3, 2007, strike on a compound owned by a local militant but used by Haqqani's fighters killed five people described as "militants" by the Pakistanis.[90]

This strike probably took place without the support of the Pakistanis, who had considered the North Waziristan–based Haqqani to be a "strategic asset" to be used in neighboring Afghanistan.[91] Although the Pakistanis were at war with the Pakistani Taliban, they still protected the Afghan Taliban. The Americans' suspicion that their Pakistani allies were working with Haqqani, their worst enemy in Afghanistan, was confirmed when the CIA intercepted communications between Pakistani ISI agents and Haqqani terrorists who subsequently carried out a suicide bombing of the Indian embassy in Kabul, which killed fifty-four people.[92]

Clearly the Pakistanis were at odds with the United States over the basing of the Afghan Haqqani Network in North Waziristan, and the CIA drones had to unilaterally carry out operations against this key Pakistani terrorist ally. (CIA drones ultimately killed two of Haqqani's sons, Mohammad and Badruddin.) The Americans and the Pakistanis were also at odds when it came to Mullah Omar and the main Afghan Taliban group he lead. Omar was allowed to live unmolested in the Pakistani town of Quetta after he told his Afghan Taliban followers not to join the Pakistani Taliban in attacking the Pakistani government. One journalist described the Afghan Taliban's sanctuary in Pakistan's Pashtun tribal zones, which included much of north Baluchistan: "As I traveled through Pakistan and particularly the Pashtun lands bordering Afghanistan, I felt as if I were moving through a Taliban spa for rehabilitation and inspiration. . . . Quetta had become a kind of free zone where strategies could be formed, funds picked up, interviews given and victories relished."[93]

Although the Pakistanis did not allow the CIA to use their drones to kill Taliban leaders in Quetta because the city was part of Pakistan proper, they did occasionally move against Afghan Taliban when pushed. This had happened on December 19, 2006, when Pakistani agents informed the Americans that a high-ranking member of the Taliban's *Shura* (Inner Council), Mullah Akhtar Usmani, was crossing into Afghanistan. Usmani was said to be in charge of Taliban operations in Afghanistan and was designated as Mullah Omar's successor should he be killed.[94] His location was determined when his telephone communications were intercepted by a drone. The CIA dispatched a Predator to his location, and the drone killed him and two of his deputies with missiles while they were driving in their car.[95]

Although ties between the CIA and ISI remained strained as the year 2007 drew to an end, such examples of occasional cooperation boded well for the uneasy Pakistani-American alliance. It became clear that further cooperation would be needed if the CIA was going to expand its drone operations further into the FATA region, which was obviously the agency's intent. Still, few could have envisioned the upsurge in killings that began in 2008.

6

The Drone War Begins

*The political consensus in support of the drone program,
its antiseptic, high-tech appeal and its secrecy have obscured
just how radical it is. For the first time in history,
a civilian intelligence agency is using robots to carry out a
military mission, selecting people for killing in a country where
the United States is not officially at war.*
—**Scott Shane,** *New York Times*, **December 3, 2009**

T here was nothing to indicate that in 2008 the CIA would transform its limited targeted assassination campaign of just a few strikes per year (between one and five) into a full-blown aerial campaign of thirty-four strikes. In just a few years the CIA had gone from having deep reservations about using drones to becoming, in the words of one agency official, "one hell of a killing machine."[1] But as the campaign stepped up, the growing perception in Pakistan was that the drones seemed to have a unique capacity to kill innocent civilians, not their actual targets, al Qaeda and Taliban militants.

A case-by-case analysis of the strikes sheds some much needed light on the nature of the drones' targets. Contrary to claims that "99 percent" of those killed in drone attacks are civilians and "1 percent" terrorists, a systematic analysis of the 2008 strikes shows that the vast majority of those killed were terrorists. Over and over again the drones seemed to find their targets with unprecedented accuracy and take them out cleanly.

THE FEDERALLY ADMINISTERED TRIBAL AGENCIES
OF PAKISTAN, 2008

The first drone strike of 2008 fit the limited HVT assassination pattern of previous years. It took place in North Waziristan and targeted an Arab, Abu Laith al Libbi, a top al Qaeda leader who directed a February 2007 suicide bombing outside a gate at Bagram Airfield, Afghanistan, that killed twenty-three people, mainly civilians, while Vice President Richard Cheney was visiting.[2] Libbi was al Qaeda's main liaison to Afghan Taliban fighters. In one of his videos he called for Westerners to be kidnapped, and in another he called for an assault on Israel.[3]

Fully aware of Libbi's growing importance to al Qaeda, American forces tried killing him in a rocket attack while he was in Afghanistan's Paktia Province in 2007, but he survived.[4] On January 29, 2008, however, his luck ran out. The notorious Libyan terrorist leader who was said to be al Qaeda's new number three was tracked down by the CIA and killed alongside twelve Arabs, Taliban fighters, and Central Asian Turkmen militants by a drone near Mir Ali in North Waziristan.[5] There were no reported civilian deaths from the strike.

Although jihadist websites around the world hailed the martyrdom of Libbi as a success, his death was perhaps the greatest loss for the terrorists since the drone assassination of the previous al Qaeda number three, Muhammad Atef, in 2001. After his death, local Taliban militants took Libbi away for burial—but only after cordoning off the destroyed compound, a practice they would continue in upcoming strikes.

The second strike of the new year took place on February 28, 2008, near the village of Kaloosha, an area in South Waziristan believed to have significant al Qaeda activity.[6] The missiles hit a house belonging to a tribesman with "well-known links to fighters in the area," according to Al Jazeera. Both Al Jazeera and the Arab newspaper the *Gulf News* reported that ten of the thirteen victims were Arabs, presumably fighters or terrorists with ties to al Qaeda.[7] There were no HVTs involved in the strike, but it was nonetheless against Arabs, which in the FATA context meant al Qaeda terrorists or fighters. Once again the strike was surgically precise, and no civilians were killed.

The third recorded strike of the year took place on March 16, 2008, in the village of Shahnawaz Kot near the town of Wana in South Waziristan. At least eighteen people were killed in this strike on a home belonging to a Noorullah Wazir, who was described by local residents as a Taliban supporter. Local residents claimed that "foreigners linked with Taliban and Al-Qaeda militants" were staying in the targeted compound at the time of the attack (thus far all the attacks in 2008 were on Arabs).[8] Although there was no proof that the attack was carried out by a drone, the *Washington Post* reported the following details based on interviews with local sources: "Local residents said they heard the sound of a warplane overhead, then three successive explosions. The strike, which demolished the house, also left several people wounded. 'When I heard the explosions, I rushed to the place where it happened. I saw dead bodies scattered everywhere,' said Aziz Ullah Wazir, a village resident. 'There were scores of people surrounding the collapsed building.'"[9] Drones often attack in waves, which could explain the multiple explosions.

By now very few people in the FATA did not know that American drones were hunting Arabs. The stepped-up pace of CIA drone strikes in the first three months of 2008 (one per month, the fastest pace by far up to this point) had finally also begun to garner considerable attention in the United States. Six days after the March 16 attack, *Newsweek* revealed that the pace of strikes had accelerated after high-level U.S. officials had reached a new understanding with President Musharraf. CIA director Michael Hayden and Director of National Intelligence Mike McConnell had convinced Musharraf that updated targeting rules were needed because the militants in the FATA were now at war with both the U.S. and Pakistani governments.[10] Hayden reportedly told Musharraf, "Mr. President, we've seen a merger. You've been slow in recognizing this merger between Al Qaeda and Pashtun extremists. Now they're coming out of the tribal areas not just to kill us, but to kill you. They're after you now."[11] Convinced that the Pakistani Taliban were now the enemy of Pakistan, Musharraf agreed to expand the parameters of the drone strikes. The new agreement gave the CIA "virtually unrestricted authority to hit targets in the border areas."[12]

In addition, a *New York Times* article mentioned a "relaxation" of the rules under which the CIA could launch strikes on al Qaeda targets. According to this article, instead of having to confirm the identity of a suspected Taliban or al Qaeda leader before attacking him with a drone, the CIA could strike convoys of vehicles that bore "the characteristics of al Qaeda or

Taliban leaders on the run."[13] Further details of this extraordinary development were revealed in a book by *New York Times* journalist David Sanger.
Sanger wrote, "In a process that had taken months, Bush had expanded what
Hayden and McConnell called 'the permissions.' He simply lowered the standard of proof needed before the Predators could strike. For the first time the
CIA no longer had to identify its target by name; now the 'signature' of a typical al Qaeda motorcade, or of a group entering a known al Qaeda safe house
was enough to authorize a strike."[14]

This new license to kill based upon mere suspicious behavior was far
beyond the original parameters of the rather limited assassination strikes
that had worried George Tenet back in 2001. A subsequent *Los Angeles Times*
article revealed even more about the expanded targeting parameters granted
to the CIA by the Bush administration in early 2008. According to the article, the CIA now had permission to rely on "pattern-of-life" analysis based
on surveillance by overhead drones that could fly for up to forty hours.[15] If
suspected militants on the ground were filmed over time engaging in suspicious activities (such as, presumably, infiltrating Afghanistan with weapons,
training in known al Qaeda and Taliban camps in the FATA, transporting
heavy weapons or explosives used in warfare against NATO or Pakistani
troops, and entering known terrorist guest houses with Arabs), then the CIA
had blanket permission to kill them based on their hostile "signatures."

Whereas previously the CIA had been limited to killing high-ranking
targets who were on a kill list known as the Joint Integrated Prioritized Target
List (this list included mainly Arab al Qaeda terrorists and high-ranking
Taliban members who were considered "personality strikes"), they now had
authorization to go after a much wider range of suspiciously acting targets,
described as rank-and-file foot soldiers, via "signature strikes." Much of the
catalyst for this new policy came from the U.S. military, which was eager to
see the CIA play a "force protection" role in disrupting cross-border insurgent
activities in neighboring Afghanistan. While the CIA was initially agnostic
about the need to go after lower-level Taliban operatives based on their pattern-of-life signatures, they soon became avid converts to this new counterinsurgency role.[16] In essence, overnight the CIA became a sort of covert
military force operating in Pakistan, only without the accountability and
scrutiny that the U.S. military faced.

The new policy essentially meant that the CIA would spread its kill net wide and start assassinating lower-level militants whose identities were not even known. The widening of what had previously been a limited, targeted assassination campaign of confirmed HVTs to include a range of suspected targets raised the hackles of some observers. Especially since the new targeting guidelines were to be based, to a certain extent, on remote-control surveillance evidence. One Bush national security aide said, "It's risky because you can make more mistakes—you can hit the wrong house, or misidentify the motorcade."[17]

There were also moral and ethical questions involved in relying on newly evolving technology to kill a wider range of people across the globe. One blogger worried, "Think about that: we're potentially killing people based not on what we know about an individual, but what we have observed solely through the camera of a drone."[18] Noah Shachtman commented in the widely read Danger Room section of Wired.com, "Once upon a time, the CIA had to know a militant's name before putting him up for a robotic targeted killing. Now, if the guy acts like a guerrilla, it's enough to call in a drone strike."[19]

While U.S. officials cautioned that the drones would be called off if there were risks of civilian casualties from a strike, the limited assassination campaign was broadening into what could best be described as an aerial war. Clearly Bush White House officials felt that under the laws of war they had the legal right to wage an asymmetric aerial campaign against terrorists and fighters preparing terrorist acts and waging war on U.S. troops from cross-border sanctuaries and havens in the autonomous tribal regions of Pakistan.

But that was not all. The demand that the CIA seek "concurrence" from Pakistan's government was later dropped from the agreement that made the CIA director "America's combatant commander in the hottest covert war in the global campaign on terror."[20] Previously the Pakistanis had been allowed the right to concur with an intended strike or to veto it. When that right was dropped from the agreement, the chances of Pakistani ISI officers with mixed loyalties warning the drone targets in advance were greatly reduced. According to former CIA officer Bruce Riedel, on several occasions the Pakistanis had tipped off drone targets in advance of a strike.[21] The most notable case of the Pakistanis tipping off a Taliban target, according to

Matthew Aid, occurred when ISI agents warned Jalaludin Haqqani of an impending drone strike.[22]

Although the expansion of the drones' targets and the end to the right of concurrences might have bothered the Pakistanis, who were always sensitive about the issues of sovereignty, collateral damage deaths, and their close ties to the Afghan Taliban (as opposed to their war with the Pakistani Taliban), larger events on the ground in the FATA at this time actually favored the widening of the campaign. In mid-December of the previous year five disparate jihadi organizations had officially organized themselves in the FATA as the TTP. They chose as their head the soon-to-be-notorious leader Baitullah Mehsud, whose assassination was outlined in chapter 1. This loose umbrella organization was created in part as a response to the drone strikes and Pakistani military incursions into the militants' de facto secessionist state.

At this time the militants also went on the offensive and conquered the last remaining free zones in Swat Valley. The suicide bombings and the invasion of Swat infuriated the Pakistani military and civilian leadership and came to be seen as a major threat, not just to the U.S. and Afghan governments, but to Pakistan itself. Many Pakistani leaders felt that the previously tolerated militants had gone too far. For this reason, the Pakistani authorities were willing to turn a blind eye to a stepped-up drone campaign against the newly aggressive Pakistani Taliban.

For its part, the CIA was better prepared than ever to take advantage of the growing hostility between the Pakistani military-government and the Pakistani Taliban thanks to the development of a deadly new drone model known as the MQ-9 Reaper. Whereas the original MQ-1 Predator was a surveillance craft that had been retroactively jerry-rigged to carry two missiles, the much larger Reaper was specifically designed as a killing platform. The $20 million Reaper could carry eight times the payload of its smaller Predator predecessor (that is, the same number of missiles as an Apache Longbow attack helicopter), was not limited to Hellfire missiles, and could deliver two five-hundred-pound Paveway laser-guided bombs or JDAMs.[23] The Reaper carried the same payload as an F-16 manned fighter jet (1.5 tons) but could stay aloft ten times longer.[24] As one military expert put it, with the Reaper "you have a lot of ammo circling overhead on call for short notice strikes."[25] The Reaper could also fly to a target three times faster than the Predator and loiter for slightly longer periods of time (up to forty-two hours). The Reaper was a Predator on steroids, and it gave the CIA what the military called "deadly persistence" in the hunt for al Qaeda. The Reaper had already made its debut on the Afghan battlefield with great effect in the fall of 2007, and it soon began making kills in Pakistan's tribal zone as well (although there were fewer of these new aircraft in operation than there were Predators).[26]

On May 14 a drone struck again, this time in the infamous Damadola region of Bajaur, a major hotbed for cross-border insurgency activity into Kunar. According to Pakistani sources, this strike took place on a compound in the hamlet of Khaza, where "militants had gathered for dinner."[27] The initial strike set off a chain of blasts from explosives collected in the targeted house. Between six and twelve people were killed in the strike and resulting explosions, which, interestingly, did not cause any uproar in the region—primarily because the compound was owned by an Afghan who was a former Taliban defense minister named Maulvi Obaidullah.

The former Taliban minster and his civilian family were not the only ones to die in the attack. At the time they were hosting an important guest named Abu Suleiman "al Jazairi" (the Algerian). Al Jazairi was al Qaeda's director of external operations and was responsible for running the terrorist group's European and British operations.[28] He was said to have trained British Muslims who traveled to Pakistan to learn how to carry out terrorist operations.[29] His death was a remarkable example of how the drones could help preempt future terrorism by killing HVTs who were plotting future attacks from remote hideouts.

A Reaper may well have been involved in the next drone strike, which occurred in the Wana region of South Waziristan on May 16. According to an al Qaeda video that eulogized those killed in the strike, the attack killed a prominent Pakistani jihadi trainer named Dr. Arshad Waheed and twenty other Taliban and al Qaeda militants.[30] Once again there were no civilian deaths on this occasion.

The next air strike, in early June, came about in more confusing circumstances and actually involved the death of Pakistani Frontier Constabulary troops. Probably delivered by a piloted U.S. aircraft, the strike took place after Afghan troops were ambushed by Taliban near the Afghan-Pakistani border. Another air strike was called in to support the troops, but it inadvertently hit a nearby Pakistani Constabulary post killing eleven Pakistani paramilitaries. Eight Taliban were killed and eleven wounded in the strike.[31] The Pakistanis were understandably furious.

This mishap was not, however, enough to damage the joint U.S.-Pakistani war on al Qaeda and the Taliban, and the drone campaign went on. Several days later a drone struck again, this time in Makeen, South Waziristan, on June 14. Pakistan news reports claimed the strike was an attempt to kill the newly appointed Pakistani Taliban leader Baitullah Mehsud.[32] No civilians were reported killed at this time.

The next drone strike took place in the village of Azam Warsak on July 27. Between five and six people, including a notorious al Qaeda weapons of mass destruction expert, Abu Khabab "al Masri" (the Egyptian), were said to have been killed in the attack.[33] Abu Khabab was described as al Qaeda's "mad scientist," and he had filmed himself killing dogs in a lab using hydrogen cyanide. the same agent used by the Nazis in their gas chambers.[34] He was also said to have been involved in making the explosives used in the terrorist bombing of the USS *Cole* and in the training of Richard Reid, the failed al Qaeda "shoe bomber" (a Brit who tried to set off a shoe bomb on a civilian-packed airliner over the Atlantic in December 2001).[35] Khabab, who had a $5 million bounty on his head, was trained to develop chemical and biological weapons to be used in mass-casualty terrorism against the West and was a major threat.[36] His death was cause for celebration among the counterterrorism experts trying to protect the United States from mass-casualty, chemical-biological terrorism.[37]

The strikes continued on August 13, 2008, with a drone attack on a militant training camp in South Waziristan run by Gulbuddin Hekmatyar, the ex-Afghan mujahideen warlord who was waging a terror campaign in Afghanistan. The strike killed a commander named Abdur Rehman and between nine and twenty-four other militants, including Turks and Arabs, according to Pakistani and U.S. sources.[38]

The barrage continued a week later on August 20 with a strike on a house in Wana, South Waziristan, that a Pakistani source described as a "known hideout for militants." Eight people, including "foreign extremists," were killed by missiles that a Pakistani official said "came from Afghanistan." Locals who were interviewed said of the wounded owner of the house, "Arabs often stayed with him."[39] No civilians were reported killed in this strike.

The number of drone strikes may have surged at this time because of the concurrent power vacuum in Pakistan; President Musharraf had been forced to resign on August 18, 2008.[40] The accelerated pace of strikes may have also come as a result of increased pressure on the White House and CIA from a recently released Government Accountability Office (GAO) report on the FATA region that found, "The United States has not met its national security goals to destroy terrorist threats and close the safe haven in Pakistan's FATA. According to U.S. officials and intelligence documents, since 2002, al Qaeda and the Taliban have used Pakistan's FATA and the border region to attack Pakistani, Afghan, as well as U.S. and coalition troops; plan and train for attacks against U.S. interests; destabilize Pakistan; and spread radical Islamist ideologies that threaten U.S. interests."[41]

As the pressure mounted on the CIA, it took advantage of the political vacuum in Pakistan, and according to *Dawn*, on August 30 it launched an attack on a house that had been rented out to "foreigners" in the Korzai area of South Waziristan. The Pakistani *Daily Times* reported that five people, including two Arabs with Canadian passports, were killed in the strike and several Punjabis were wounded.[42] By this time Waziristan was home to hundreds, if not thousands, of Punjabi militants who had gradually become known as the "Punjabi Taliban." At their base in Waziristan, the Punjabi Taliban began to train suicide bombers for missions throughout Punjab and the rest of Pakistan.

One aspect of preceding strike accounts deserves attention, namely, the fact that most of the details on the strikes (such as claims that the owner of a targeted house "rented out to foreigners") came from Pakistani journalists interviewing locals at the scene. Over and over again Pakistani sources described the attacks as drone strikes on houses or compounds with ties to the Taliban and al Qaeda foreigners. Note that these same Pakistani sources did not describe the targets of the strikes as "innocent civilian residences," even though Pakistani journalists tend to display anti-American attitudes. According to one study carried out by the Pakistani newspaper *Dawn*, a full 67 percent of Pakistani journalists interviewed in an opinion poll found the drone strikes to be "acts of terrorism."[43] Yet the Pakistani media sources seemed remarkably frank in providing details that would indicate that the targets of the drone strikes were invariably linked directly or indirectly to al Qaeda terrorism or, less often, to Taliban terrorist-insurgency activities.

On the following day, August 31, a drone hit a house near Miram Shah in North Waziristan. Once again Pakistani sources reported that several "foreigners" were killed in the strike, but so were a woman and a child, the third pair of civilian casualties that year.[44] By this time the drone strikes had become such an accepted part of the battle rhythm of the FATA zone that they did not cause much of an uproar in Pakistan. In my interviews with Pakistanis in the tribal zones and in parts of Pakistan proper in the summer of 2010, I found that this apathy was a product of two unique circumstances. First, most Pakistanis considered the autonomous FATA region to be removed from the rest of Pakistan proper, like a U.S. territory, such as Samoa, Guam, or Puerto Rico, or the Wild West in the 1800s. Second, U.S. combatants or pilots were not directly involved in the assassinations, which targeted people the locals knew were involved in militancy or terrorism. Because unmanned "robot" planes carried out the strikes, the CIA's violation of Pakistani sovereignty was somehow more palatable than it would have been had they been carried out by manned bombers, like the Soviet air strikes of the 1980s.

Although the powerful pro-Taliban Islamist parties occasionally criticized the strikes—and even those on the Pakistani secular left who despised the Taliban fundamentalists criticized the strikes as "acts of imperialist aggression"—there were no mass protests like those following the Chenagai and Damadola strikes. Seemingly, by the summer of 2008 the Pakistani leadership had accepted the drone strikes in the FATA as an unfortunate but necessary evil in the state's new war on the Taliban. If Pakistan wanted billions of dollars in U.S. aid, it had to allow the CIA to kill the common enemy by using its drones. But there would be no U.S. ground forces in the FATA. The hunt would be limited to remote-control planes.

This last point was challenged on September 3, 2008, when the United States shocked Pakistan by launching a helicopter-borne special operations raid on the village of Musa Nika in the Angoor Ada region of South Waziristan. The village of Musa Nika was a well-known cross-border sanctuary for

Taliban insurgents who infiltrated into Afghanistan and attacked U.S. and Coalition troops.[45]

According to the *New York Times*, just prior to the September 3 night raid, U.S. and Afghan troops had pursued a group of Taliban fighters that had attacked them in Afghanistan. The enemy escaped by retreating across the border to the village of Musa Nika.[46] There they were safe in their sanctuary— until 3:00 a.m. on September 3. At that time three to five Black Hawk helicopters roared over the village and began to disgorge U.S. Navy SEALs into the target. In the ensuing mayhem, between nine and twenty people were killed, including women and children. The troops in the helicopters killed or captured several suspected Taliban militants and then disappeared into the night. The attack lasted no more than an hour.

The following day U.S. officials sounded upbeat and hinted that the raid might be a sign of things to come.[47] One U.S. military officer said, "You can't allow a haven. You have to get the areas that they rest, relax and train."[48] A senior U.S. official commented, "The situation in the tribal areas is not tolerable. We have to be more assertive. Orders have been issued." It later emerged that President George Bush himself had approved the orders calling for a new policy of conducting raids in Pakistan without notifying the Pakistanis.[49]

As word spread in Pakistan that Pakistani women and children had been killed by U.S. troops on Pakistani soil, there were howls of outrage from Pakistani leaders across the board. One senior Pakistani official called the night raid a "cowboy action" and criticized it for not targeting anyone "big."[50] The Pakistani Foreign Ministry condemned the attack, calling it "unacceptable" and "a grave provocation . . . which has resulted in immense loss of civilian life."[51] The following day the Pakistani parliament passed a resolution condemning the raid, and the foreign minister told the National Assembly, "There is no high-value target or known terrorist among the dead. . . . Only innocent civilians, including women and children, have been targeted."[52] U.S. Ambassador Anne Patterson was also summoned to hear a "very strong protest" at the Foreign Ministry.

The Pakistani military raised the biggest objection. Using bellicose terms more suited for an enemy than an ally, the new head of Pakistan's army, Gen. Pervez Kayani, said that Pakistan's territorial integrity would be "defended at all cost" and that "reckless actions only help the militants and further fuel the militancy in the area."[53] Lest there be any ambiguity, Kayani added, "There is no question of any agreement or understanding with the coalition forces whereby they are allowed to conduct operations on our side of the border. . . . No external force is allowed to conduct operations inside Pakistan."[54]

Although the Pakistanis were willing to countenance the occasional civilian death or attacks on militants if they were administered by unmanned drones, U.S. troops landing on Pakistani territory was essentially construed as an act of war. Joint Special Operations Command (JSOC) had grand dreams of commencing a "hot pursuit" targeting campaign against Taliban insurgents operating from safe havens in Pakistan, but clearly the hunt would have to be left to the unmanned drones if the Pakistanis were to be placated.

In fact a drone strike was launched the following day, September 4. The strike was again in North Waziristan, this time hitting a house whose owner was "known to host foreigners," according to locals who spoke to the Agence France-Presse (AFP) news service.[55] Four people were killed in that strike on territory said to be under the control of Jalaludin Haqqani. None of the slain were reported to be civilians.

A drone launched another attack one day later on a village near the Afghan border of North Waziristan known as Al Must. One report claimed that six to twelve people were killed in the strike, including "men of Arab descent," two women, and three children. According to a local source, "Three missiles hit the two compounds, which he said belong to two residents of Al Must, Hakeem Khan and Arsala Khan. It is common for families in these areas to rent part of their compound to foreigners, especially Arabs who are involved in planning attacks against NATO forces in Afghanistan, residents said."[56] Once again the rule seemed to be that Arabs were lightning rods for drone strikes and those who rented rooms to them or were related to them ran the real risk of being killed by a Predator or Reaper.

The next strike, on September 8, 2008, was against the aforementioned Afghan Taliban leader Jalaludin Haqqani, whose terrorist insurgents had killed more than fifty people in a suicide attack on the Indian embassy in Kabul just two months earlier. The drone strike targeted a madrassa in the

North Waziristan town of Dande Darpakhel, which was the Haqqani Network's Pakistani headquarters. Although the strike aimed to kill Jalaludin Haqqani, or his son Sirajuddin, who was gradually taking control of operations from his aging father, it missed both men. Instead twenty-three people, including one of Jalaludin's two wives, his sister, his sister-in-law, and eight of his grandchildren, were killed.[57] Four al Qaeda operatives were also said to have been killed in the attack.[58] A large number of civilians were killed in this attack, which raises the question, Why would Haqqani house al Qaeda Arabs who could attract drone attacks near his wife and grandchildren? Also, did the CIA know there were civilians present at the time and decide to carry out the attack with the aim of killing the notorious Haqqani, despite the risk of civilian casualties?

Another strike took place in a suburb of Miranshah just four days later on September 12, 2008.[59] One source reported twelve "rebel fighters" killed in the strike, whereas another source claimed "seven Taliban" were killed.[60] Reports from the strike on two buildings said that once again it appeared that civilians had been killed as well.[61]

Five days later a drone strike killed Abu Ubaydah al Tunisi, a Tunisian al Qaeda leader, and between four and six other men in the Bangar Cheena region of South Waziristan. There were no reports of civilian deaths on this occasion, and *Dawn* claimed that the Tunisian leader and his men were "delivering rockets to a militant camp near the Afghan border."[62] This appeared to be a clean strike on a person who could be described as an enemy combatant.

That same day, September 17, 2008, the Pakistani news site *Geo.TV* reported, "The Pentagon has claimed that U.S. led coalition forces carried out another drone strike on an ammunition storage facility of Taliban, in which one al-Qaeda member and 3 Taliban militants were killed. The U.S. authorities said they shared the news with Pakistani officials after conducting the strike."[63] This intriguing Pentagon statement can be interpreted in several ways. One interpretation is that the Pentagon (and not the CIA) chose to inform the Pakistanis of a military drone strike on Taliban and al Qaeda that occurred inside Afghanistan. This alternative seems strange, however, for the U.S. military had no reason to share with Islamabad an after-action report on a rather limited drone engagement in a neighboring country. The Pentagon report of a drone strike is also strange in that the military is primarily allowed to operate in recognized combat zones, but not in civilian areas in a country where the United States is not officially at war. The other interpretation is that the U.S. military launched a cross-border drone strike on an undisclosed location in Pakistan in what can best be described as a force protection role and then informed the Pakistanis. This scenario in which the U.S. military attacked an ammunition dump in Pakistan seems more plausible as the CIA had, and still has, a policy of not officially discussing its individual drone strikes.

This second scenario seems even more plausible in light of writer Jeremy Scahill's revelations about the existence of a limited U.S. military drone campaign in Pakistan. In a piece for the *Nation* he described "a secret US military drone bombing campaign that runs parallel to the well-documented CIA predator strikes." One insider military source speaking of the JSOC drone campaign described it as "a parallel operation to the CIA" and called the CIA and JSOC campaigns "two separate beasts."[64] Another referred to the military campaign as "a separate fleet of U.S. drones operated by the Defense Department [that] will be free for the first time to venture beyond the Afghan border under the direction of Pakistani military officials."[65]

Scahill further reported, "In 2006, the United States and Pakistan struck a deal that authorized JSOC to enter Pakistan to hunt Osama bin Laden with the understanding that Pakistan would deny it had given permission. Officially, the United States is not supposed to have any active military operations in the country."[66] According to one account, JSOC Navy SEAL teams had raided Pakistan as many as twelve times before raids were stopped after the assault on Angoor Ada in September 2008.[67] Following this public relations fiasco, there were several instances in which Taliban fighters who had attacked U.S. troops were chased by U.S. airpower into Pakistan and killed. In the most notable of such incidents, fifty Haqqani Network fighters were killed after fleeing into Pakistan.[68]

Further evidence of an ultrasecret drone campaign led by Gen. Stanley McChrystal's JSOC (McChrystal later became head of all NATO forces in Afghanistan) came from Noah Shachtman writing for Wired.com. During a visit to a secret U.S. military base (most likely in Karshi-Khanabad, or K2, Uzbekistan), he reported,

Today, those [cross-border JSOC drone] missions have become a regular occurrence. The U.S. Air Force has a fleet of Predator and heavily armed Reaper drones, stationed at Kandahar and Jalalabad Air Fields in Afghanistan. All of these robotic aircraft are allowed to venture occasionally into Pakistani airspace to pursue militants. The government in Islamabad just has to be notified first. Some of the Predators also fly into Pakistan on operations in conjunction with or in support of Islamabad's military.

These missions are remotely flown by U.S. Air Force pilots at Creech Air Force Base, Nevada; the footage is shared with the Pakistani government, including at joint coordination centers on the border.

In addition, some of the military's Predators and Reapers are placed under the operational control of the CIA, which uses them to conduct their own strike and surveillance missions. Some of those drones take off from Jalalabad, others from within Pakistan itself, at a remote base called Shamshi. According to the *New York Times*, those aircraft are operated out of CIA headquarters in Langley, Virginia. . . . From what I can tell, these CIA missions comprise the bulk of the drone flights over Pakistan. And the military has, at times, encouraged the notion that operating the unmanned aircraft was the spy agency's job.[69]

Regardless of whether the source of the upswing in drone strikes in September 2008 was the military's JSOC or the CIA, there were few criticisms from the Pakistani authorities, who were caught up in the election of their new president, Ali Asaf al Zardari. Zardari was the widower of the recently slain presidential candidate Benazir Bhutto, who had been killed by a Baitullah Mehsud Taliban suicide bomber upon her return to Pakistan in 2007. For this reason the Americans hoped Zardari would be a strong ally, and in this expectation they were not disappointed. Zardari seemed to be willing to stand by the Americans and the war on the terrorists who threatened his state, even if it cost him some popularity among his own people. Zardari referred to the Taliban as a "cancerous" threat to Pakistan and told the Americans that he would "take the heat" if the United States launched a cross-border raid to capture an HVT like bin Laden or Zawahiri.[70]

On September 30 the CIA took advantage of the new climate of cooperation, and a drone struck again, this time in Mir Ali, North Waziristan. The drone missile struck the house of a "local Taliban commander" and killed six people.[71] On the night of October 3–4 a drone struck yet again. A senior Pakistani military official said of the attack, "Our reports suggest that around 20 suspected militants were killed when a missile hit a house in Mohammad Khel village in North Waziristan. Most were foreigners."[72] There were no recorded civilian deaths on this occasion.

On October 9 a house east of Miranshah, North Waziristan, whose owner was hosting foreigners, was hit by a drone. At least six people, including three Arabs, were killed in the attack.[73] Three days later a drone struck again in Miranshah, killing five people, but there were no reports on the victims' identities.[74] Just four days later, on October 16, a drone struck in the village of Taparghai, South Waziristan. This strike killed four people, "some of them Arabs," including Khalid Habib, the number four in al Qaeda and head of their operations in the Pakistani tribal regions.[75] Habib and the other Arabs were killed in their Toyota station wagon in one of the first recorded hits on a vehicle in Pakistan.

On the night of October 22–23 a drone struck again, this time against the Haqqani Network in a village just outside Miranshah, North Waziristan. Locals reported, "One missile hit a room in the compound where the militants were sleeping."[76] Three days later a drone struck a "facility/alleged militant compound" in a village near Wana, South Waziristan.[77] Twenty people were killed in this strike, among them Mohammad Omar, a Taliban commander who had been close to Nek Muhammad.[78] Interestingly enough, the drone missile completely destroyed Omar's compound but only "damaged" two neighboring houses.

Five days later, on October 31, a drone struck again in Wana, killing six foreigners and a local tribesman who was hosting them.[79] That same day a drone also struck in neighboring North Waziristan, killing a prominent al Qaeda leader named Abu Jihad al Masri and two other "rebels" who were traveling in a car with him. Masri was an Egyptian and a high-ranking propaganda expert who appeared in a video with Ayman al Zawahiri.[80] There were no civilian bystander casualties in this strike.

Shortly after these strikes, on November 4, the Americans chose a new president, Barack Obama. But this did not slow down the pace of the kills in

Bush's final months in office. On November 7 a drone attacked an al Qaeda training camp located near the Afghan border in the village of Kumsham, North Waziristan. According to a Pakistani source, "Between 11 and 14 militants, mainly foreigners, were killed in the strike."[81] Among those killed were seven al Qaeda operatives and one Taliban commander. Once again there were no reports of civilian casualties in the attack.

On November 14 a drone attacked again, this time in the village of Garyom near Miranshah, North Waziristan, an area described as a "hotbed of Al Qaeda and Taliban support."[82] In this strike twelve people, including "nine foreign militants, believed to be Al Qaeda fighters," the homeowner, and two of his family members, were also killed.[83] The message for other Pashtuns who might have been tempted to host foreign al Qaeda fighters in their houses was clear: You run the risk of having your home and family destroyed if you provide sanctuary to foreign terrorists-militants.

At this time President Zardari, who was visiting New York, publicly claimed that the drone strikes were "counter-productive and violated Pakistan's sovereignty."[84] This pro forma statement was obviously meant to garner support among his own people, who strongly disliked the idea of a foreign power operating with impunity on their own soil, killing what many believed were almost exclusively innocent Pakistani citizens. The Pakistani people wanted their leaders to publicly stand up to the American "invaders." But secretly Zardari was said to have told the Americans, "Kill the seniors. Collateral damage worries you Americans. It does not worry me."[85] He also said, "There are no differences between Pakistan and the US over any issue, including drone attacks."[86] He even made a plea for the United States to give his country access to the drones. He told a U.S. delegation, "Give me the drones so my forces can take out the militants." That way, "we cannot be criticized by the media or anyone else for actions our Army takes to protect our sovereignty."[87]

Zardari seemed to appreciate that the drone attacks were helping his country avoid military casualties they would have sustained had they directly attacked the terrorists' lairs. According to a Wikileak cable from Ambassador Patterson, "Referring to a recent drone strike in the tribal area that killed 60 militants, Zardari reported that his military aide believed a Pakistani operation to take out this site would have resulted in the deaths of over 60 Pakistani soldiers."[88] Similarly, a spokeswoman for Zardari's political party, the Pakistan People's Party, declared, "There is a segment in the country who support the drone attacks, and they feel that drone attacks have been helpful in eliminating many of the militants."[89] One military officer told AFP, "The Pakistani army supports drone strikes because they are efficient for eliminating TTP people . . . and give it a good reason not to start a dangerous offensive in North Waziristan."[90] Pakistan's ambassador to the United States added, "Pakistan has never said that we do not like the elimination of terrorists through predator drones."[91]

The Wikileaks documents from 2009 and 2010 show that Pakistani prime minister Yousuf Gilani similarly opined of the drones in private, "I don't care if they do it as long as they get the right people. We'll protest in the National Assembly and then ignore it."[92] Gen. Shah Shuja Pasha summed up his views of the Taliban and al Qaeda when he said, "We would obviously like to fix these rogues. They are killing our own people and are certainly not the friends of this country."[93] In addition, General Kayani asked the United States for "continuous Predator coverage of the conflict area" during his forces' campaigns against the Taliban in the FATA.[94] This request was answered in the affirmative during Pakistani operations in South Waziristan, and one U.S. military official told the *Los Angeles Times*, "We are coordinating with the Pakistanis. And we do provide Predator support when requested."[95]

Kayani's public position was quite different, however, according to a classified U.S. cable about Kayani and the government of Pakistan (GOP) released by Wikileaks:

> The strikes have put increasing political pressure on the Pakistani government, which has struggled to explain why it is allowing an ally to violate its sovereignty. The GOP so far has denied recent media reports alleging that the U.S. is launching the strikes from bases in Pakistan. Kayani knows full well that the strikes have been precise, creating few civilian casualties, and targeted primarily at foreign fighters in the Waziristans. He will argue, however, that they undermine his campaign plan, which is to keep the Waziristans quiet until the Army is capable of attacking Baitullah Mehsud and other militants entrenched there.[96]

In his book *The Most Dangerous Place: Pakistan's Lawless Frontier*, Pakistani journalist Imtiaz Gul wrote of a similar disconnect between what the Pakistani leadership secretly wanted and their public stance. Gul wrote,

> Most Pakistanis, including members of the media and mainstream political leaders, view the attacks as a violation of their national sovereignty. But privately even top generals support drone strikes. In a recent meeting with a handful of Pakistani journalists, a very senior general told us, "As long as they take out the guys who are a threat to us all, why crib about it?" Leading government officials, including Prime Minister Gilani, will agree even if publicly they condemn the drone strikes.[97]

One former U.S. official said of the Pakistanis' formulaic criticisms of the drone strikes, "There's always been a double game. There's the game they'll play out in public, but there has always been good cooperation."[98] National security analysts Peter Bergen and Katherine Tiedemann at the Washington, D.C.–based New America Foundation summed up Pakistan's duplicity:

> For Pakistani politicians, the drone program is a dream come true. They get to posture to their constituents about the perfidious Americans even as they reap the benefits from the U.S. strikes. They are well-aware that neither the Pakistani Army's ineffective military operations nor the various peace agreements with the militants have done anything to halt the steady Talibanization of their country, while the U.S. drones are the one surefire way to put significant pressure on the leaders of the Taliban and Al Qaeda. This is called getting to have your chapati and eat it too.[99]

Some Americans who knew what was going on vis-à-vis Pakistan's public criticisms of the drone strikes found Islamabad's position to be hypocritical. Senator Carl Levin, chair of the Senate Armed Services Committee, for example, summed up U.S. frustrations with the Pakistanis' double-dealing when he said, "For them to look the other way, or to give us the green light privately, and then to attack us publicly leaves us, it seems to me, at a very severe disadvantage and loss with the Pakistani people."[100] But this seemed to be the price the U.S. government was willing to pay to launch the drone attacks that the Pakistani government felt it could only secretly support.

The next strike was on November 19. It was very unusual in that it did not take place in the FATA; up until this point, all drone strikes in Pakistan had taken place in North and South Waziristan and, to a much lesser extent, Bajaur. The November 19 strike took place in the village of Jani Khel, in the province of Bannu, which is located in the North-West Frontier Province, a Pashtun-dominated "settled" land that was one of Pakistan's four main provinces. (This constituent part of Pakistan became known as Khyber Pakhtunkhwa in 2009.) Three "foreigners" and one local Taliban fighter were killed in the strike on a compound run by a Taliban leader named Parpand.[101] No civilian deaths were recorded. The Pakistani Foreign Ministry nonetheless lodged a protest with Ambassador Patterson owing to the depth of the strike into Pakistan proper.[102]

The next attack took place three days later, on November 22, in the village of Ali Khel in North Waziristan. At least four people were killed in the strike, among them a very interesting British-Pakistani terrorist named Rashid Rauf. Rauf had fled his native Britain and moved to Pakistan after killing his uncle and escaping an arrest warrant for his involvement in the notorious plot to use liquid explosive to blow up ten civilian jet airliners in 2006.[103] Rauf was arrested in Pakistan, but before he could be put on trial, he escaped. He subsequently married the daughter of the founder of Jaish al Muhammad, a Pakistani militant group, and began to plan other terrorist strikes. Justice finally caught up with the murderer/al Qaeda terrorist when the drone killed him in North Waziristan. Rauf's Pakistani lawyer played on local Pakistani sentiments when he said, "He was an innocent man a god-fearing, devout polite man and this is an extra-judicial killing."[104]

The next strike took place on November 29 in Chashma, North Waziristan, and killed three people. Information about the victims' identities is not available.[105] There was an uncharacteristic lull in drone attacks until one struck an undisclosed location in South Waziristan on December 11, killing seven Punjabi militants.[106] There were no recorded civilian casualties in this strike.

The penultimate strike of the year occurred in the village of Tapi Tool in North Waziristan and killed two people. No information about the victims' identities is available.[107] On December 22, the drone blitz of 2008 ended with a crescendo with two separate strikes on Taliban vehicles in the villages of Karikot and Shin Warsak. Eight "militants," including fighters who fired on a drone with a truck-mounted antiaircraft gun, were killed in the strikes.[108] Thus the year's campaign ended with what Bergen and Tiedemann call "a legacy-building effort to dismantle the entire Al Qaeda top leadership."[109]

The questions that remain at this point are, How did the campaign of 2008 break down in numbers of slain civilians versus militants and where did the strikes take place? A case-by-case analysis of the strikes (each of which is documented by Pakistani or Western sources of repute) leads to the following startling conclusion. The 2008 drone campaign in Pakistan resulted in 317 deaths. Of these, 249 were confirmed militants or terrorists, 36 were classified as unknown, and 32 were confirmed as civilians. In other words, according to the available sources (a majority of which were Pakistani), approximately 10 percent of the victims of the 2008 drone bombing campaign were confirmed as civilians. Although this percentage is higher than desirable, it is also a refutation of the claims that the vast majority of those who die in the drone strikes are civilians.[110]

As for the militants, they made up the vast majority of the drone victims. Surprisingly, 162 of the 249 slain militants/terrorists were not Pakistanis; they were instead Arabs, Turks, Central Asians, Canadians, Brits, and Afghans. This important distinction has not been made before and points to a real effort by the CIA in 2008 to kill foreign terrorists instead of local Pakistani Taliban militants. The locals in the FATA had to realize that the drones were not "killing Pakistani civilians every day" (in fact the thirty-two civilians were killed on seven different days in 2008). On the contrary, the CIA was clearly able to distinguish between foreigners who were linked to al Qaeda and local Taliban and to surgically target the former at a much higher rate than has previously been disclosed.

The numbers suggest that the CIA had an effective on-the-ground spy program whereby it was—and is—able to use local informants to track foreign terrorists and pinpoint the *hujras*, madrassas, training camps, and vehicles they were in for precisely targeted destruction. In most cases when civilian women, children, or local sympathizers died, they were in close proximity to a targeted terrorist or militant. Many, if not most, of the slain civilians, such as Haqqani's wife and grandchildren or the wife and children of the Afghan Taliban member Maulvi Obaidullah, were related to terrorists or insurgents. One can assume that if they were adults, they were aiding and abetting their family members. In other words, in the 2008 blitz, if you were not related to a terrorist, involved in harboring terrorists, or involved in terrorist or militant activities, the odds that you would be killed by a drone in the FATA (or the one strike in Bannu) were very slim. Once again, this fact had to be well known among people living in the FATA.

In geographic terms, of the thirty-four strikes that occurred in 2008, sixteen were in South Waziristan; fifteen in North Waziristan; one in Bajaur Agency; one in Bannu, which was outside of the FATA; and one in an undisclosed location in the FATA. With the exception of Bannu, all the strikes occurred in FATA territory that was part of a de facto secessionist state run by various Taliban warlords.

In summary, in 2008 the CIA was largely killing non-Pakistanis and, to a lesser degree, local Taliban in a secessionist part of Pakistan that was openly at war with both Pakistan and the Afghan government. What is remarkable, especially in light of the small number of civilian deaths, is that neither the CIA nor any other branch of the U.S. government sought to disabuse the Pakistanis or Westerners of the misguided notion that the drones were invading Pakistan from Afghanistan and killing mass numbers of civilians. Far from making a case-by-case defense of the campaign using Pakistani sources, the CIA stubbornly refused to comment on the drone war.

Without an American public relations campaign to counteract the critics' attacks on the drone efforts, they remained a mystery for most outsiders, who assumed the worst. But in 2009 and 2010 new light began to be shed on the drone strikes from a variety of sources, including U.S. senators, Pakistani politicians, journalists, American-based scholars, human rights activists, pollsters, Google Earth, and even the people of the FATA themselves. Although these revelations did little to change the conventional wisdom that drones invaded Pakistani airspace and indiscriminately killed Pakistani civilians at random, they began to shed some new light on the murky drone campaign for those who cared to dig deeper.

On the eve of these developments, Barack Obama took office on January 20, 2009. He let it be known that he would discontinue many of the more controversial antiterrorism practices of the Bush administration, such as water-boarding interrogations and the rendition of prisoners to CIA "black sites." He also promised to close the offshore prison camp at Guantánamo Bay and try its prisoners in the U.S. judicial system. Obama clearly saw these Bush-era practices as public relations disasters in perhaps the most important war with Islamist extremists: the war for the hearts and minds of millions of Muslims. He aimed to create a new dialogue with Muslims. In his first major international speech, which was delivered in Cairo, Egypt, Obama promised a new era of understanding toward the Muslim world. He promised to respect Islam and end the distrust between the United States and Muslims that had been exacerbated in particular by the bloody U.S. invasion of Iraq.

In light of these promises, many antiwar activists in the West and antidrone voices in Pakistan felt confident that Obama would have a different take on the drone assassination campaign than his predecessor. The question on the minds of many—from the halls of power in Washington to the Taliban *hujras* in Waziristan to the Pakistani military headquarters in Rwalpindi—was, would the newly elected Democratic president continue the drone policies of his Republican predecessor?

A Predator drone armed with Hellfire missiles over Afghanistan. *(U.S. Air Force photo by Leslie Pratt.)*

Reaper drone in hangar. *(U.S. Air Force photo by Erik Gudmondson.)*

Loading a Hellfire missile onto a Predator drone. *(U.S. Air Force photo by Sabrina Johnson.)*

A Predator with bomb images painted on its side to symbolize Hellfire missiles fired in combat. *(U.S. Air Force photo by Larry E. Reid Jr.)*

A Reaper drone with ground crew in Afghanistan. *(U.S. Air Force photo.)*

Predator pilots flying a drone over Iraq. *(U.S. Air Force photo by Steve Horton.)*

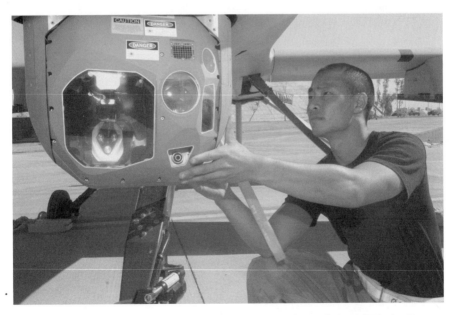

The high-resolution camera and sensor ball on a Predator drone. *(U.S. Air Force photo by Cohen Young.)*

Loading a Hellfire missile onto a Reaper drone. *(U.S. Air Force photo by Lance Cheung.)*

LEFT: A Predator armed with Hellfire missiles. *(U.S. Air Force photo.)*

BELOW: Frontal view of a Reaper armed with Hellfire missiles and Paveway guided bombs. *(U.S. Air Force photo by Brian Ferguson.)*

A Reaper armed with Hellfire missiles. *(U.S. Department of Defense photo.)*

A Reaper drone with GBU-49 GPS-guided bombs. *(U.S. Air Force photo.)*

A Reaper armed with Hellfire missiles and bombs. *(U.S. Air Force photo by Staff Sgt. Brian Ferguson.)*

A Reaper drone. *(U.S. Air Force photo by Kristi Machado.)*

A Predator at Bagram Air Base, Afghanistan. *(U.S. Air Force photo by Demetrius Lester.)*

Drone pilots. *(U.S. Air Force photo by James L. Harper Jr.)*

Side view of a Reaper. *(U.S. Air Force photo by Larry E. Reid Jr.)*

A Predator with an F-16 in the background. *(U.S. Air Force photo by Shannon Collins.)*

A Predator taking off on a mission. *(U.S. Air Force photo by Brian Ferguson.)*

7

Who Is Being Killed in the Drone Strikes?

It is legitimate to target the people who are targeting you.
—**Democratic senator Carl Levin**

This is a targeted, focused effort at people who are on a list of active terrorists who are trying to go in and harm Americans, hit American facilities and bases.
—**President Barack Obama**

hree days after Barack Obama's inauguration, the world learned the new president's views on the CIA's assassination campaign in FATA when two separate drone attacks were launched, one in South Waziristan and one in North Waziristan. A *Washington Post* headline the next day read, "2 U.S. Airstrikes Offer Concrete Sign of Obama's Pakistan Policy."[1] These strikes encapsulated all the benefits and pitfalls of the drone campaign. Whereas the first one was clean and killed ten militants, among them seven "foreigners," the second strike, on a *hujra* owned by a Taliban militant, killed several of the Taliban target's relatives, including children.[2]

This development disappointed those who were opposed to the drone campaign. Several months later one of those disappointed Americans created an iconic image of the antidrone movement by photoshopping an image of the Nobel Peace Prize onto the nose of a drone firing a missile. (Obama won the Nobel Peace Prize in 2009.) In the tribal regions of Pakistan a group of antidrone protestors organized by the Jamaat e Islami (the Islamist Party) held aloft banners that read, "Bombing on Tribes. Obama's First Gift to Pakistan," and the Taliban were said to have drummed up "more hatred for President Obama than for President Bush."[3] By 2011 a Pakistani newspaper was describing the new American president as "the Rambo of Drone Warfare."[4]

Far from ending the drone campaign in Pakistan, Obama ratcheted it up to unprecedented levels: there were 54 strikes in 2009, 117 in 2010, 64 in 2011, and 46 in 2012. In those four years Obama ordered roughly 281 drone strikes.[5] In fact, the Obama administration ordered drone strikes once every four days on average, compared to the Bush administration, which ordered a drone strike every forty days on average.[6] Obama ultimately expanded the CIA drone program to fourteen "orbits" (an orbit consisted of three drones).[7] For all intents and purposes, Obama doubled down on the policy of assassinating terrorists in Pakistan without much criticism from his own party or from Republicans, whose president had begun the drone campaign. By the fall of 2012 Obama had, without any real opposition, carried out 283 strikes in Pakistan (six times more than Bush during his eight years in office).[8]

Obama appeared to be more focused on winning the war in Afghanistan than Bush was. Soon after coming to power he announced that he would send an additional thirty thousand troops to fight in what had become known as "the Forgotten War." During his speech announcing the troop surge in Afghanistan, Obama made his intentions toward the Taliban and al Qaeda abundantly clear: "We are in Afghanistan to prevent a cancer from spreading throughout the country. But this same cancer has also taken root in the border region of Pakistan. We cannot tolerate a safe haven for terrorists whose location is known and whose intentions are clear."[9]

Obama also broadcast his intentions when he announced, "I will not hesitate to use force to take out terrorists who pose a direct threat to America. . . . I will ensure that the military becomes more stealthy, agile, and lethal in its ability to kill terrorists."[10] That Obama felt this way should have come as no surprise for those who followed his speeches on the campaign trail when he was running for president. In 2007 and 2008 Senator Obama made it clear to voters that should he become president, he would refocus U.S. energy from Iraq, which he called a "war of choice," to Afghanistan, which he called a "war of necessity." In an August 2007 speech Obama firmly declared, "I understand that President Musharraf has his own challenges. But let me make this clear. There are terrorists holed up in those mountains who murdered 3,000 Americans. They are plotting to strike again. It was a terrible mistake to fail to act when we had a chance to take out an al Qaeda leadership meeting in 2005. If we have actionable intelligence about high-value terrorist targets and President Musharraf won't act, we will." Lest there was any ambiguity about his stance, Obama continued,

> We have a difficult situation in Pakistan. I believe that part of the rea-son we have a difficult situation is because we made a bad judgment going into Iraq in the first place when we hadn't finished the job of hunting down bin Laden and crushing al Qaeda. So what happened was we got distracted, we diverted resources, and ultimately bin Laden escaped, set up base camps in the mountains of Pakistan in the north-west provinces there.
>
> They are now raiding our troops in Afghanistan, destabilizing the situation. They're stronger now than at any time since 2001. And that's why I think it's so important for us to reverse course because that's the central front on terrorism. They are plotting to kill Americans right now. As Secretary Gates, the Defense secretary, said, the war against terrorism began in that region, and that's where it will end.[11]

By personally condoning the drone strikes, President Obama was sim-ply fulfilling his campaign vow to refocus on Afghanistan and go after ter-rorist targets hiding out in Pakistan. Obama appeared set to rely on the Bush-era congressional resolution that authorized the president to "use all necessary and appropriate force against those nations, organizations, or per-sons" found to be linked to the 9/11 attacks. By year's end more than five hundred militants and a far smaller number of civilians living in the FATA region would be killed as the aerial assassination campaign that was begun under Bush gained momentum under Obama.

FEDERALLY ADMINISTERED TRIBAL AGENCIES, 2004–2013

Space will not allow an in-depth, case-by-case analysis of each of the drone strikes under President Obama, but the trends reflected in 2008 were repeated in 2009, 2010, 2011, and 2012 with only a few exceptions (such as more attacks in North Waziristan and a tendency to go after more mid- or low-level Pakistani and Afghan Taliban targets instead of high-value foreign al Qaeda targets). Only those drone strikes that killed high-ranking al Qaeda or Taliban leaders will be analyzed here in detail. This analysis of the HVTs killed in Pakistan in 2009 sheds a fascinating light on the more notable successes of the drone campaign.

The campaign of 2009 began on January 1, when five militants were killed in a strike in South Waziristan. Among those killed in the strike were a high-ranking Kenyan al Qaeda explosives expert named Usama al Kini and his lieutenant Sheikh Ahmed Salim Swedan. Kini had trained the terrorists who carried out the bombings of the U.S. embassies in Kenya and Tanzania that killed hundreds of primarily African civilians in 1998. Considering the blood on this mass murderer's hands, much of it Pakistani, his death was celebrated in the United States and Pakistan, as well as Africa.[12]

Other militant leaders killed soon thereafter included Abdullah Hamas al Filistini, a Palestinian al Qaeda trainer; Khwaz ali Mehsud, a senior deputy to Baitullah Mehsud; Mufti Noor Wali, a suicide bomb trainer for both al Qaeda and the Taliban; Kiyafetullah Anikhel, another of Baitullah Mehsud's commanders; and finally on August 5, 2009, the most-wanted man in Pakistan, Baitullah Mehsud himself.[13] His death was quietly celebrated throughout Pakistan by those moderates who abhorred the growing influence of the extremists.

Twenty-two days later, on August 27, a drone struck another interesting terrorist, named Tahir Yuldushev. Yuldushev was the most-wanted man in his native Uzbekistan. There he was wanted for founding the Islamic Movement of Uzbekistan (IMU), a terrorist group that carried out a bloody bombing campaign in that country that killed numerous civilians.[14] The IMU also kidnapped Americans in the region and spread terror to neighboring Kyrgyzstan and Tajikistan, killing scores.[15] On August 27, 2009, the Pakistanis, however, announced that a U.S. drone attack on a group of Uzbeks had killed the notorious Yuldushev. (His successor, Uthman Adil, would subsequently be killed by a drone in 2012.)[16]

On September 7, 2009, another big fish was taken out in South Waziristan: Mustafa al Jaziri, an Algerian who was a member of al Qaeda's *shura.* The following day Maulvi Ismail Khan, a Haqqani commander, was killed in North Waziristan. Six days leader Najmuddin Jalalov, the founder of a splinter group of Uzbekistani terrorists known as the Islamic Jihad Group, was killed in a drone strike. Jalalov's Islamic Jihad Group became well known in Germany after three of its members were arrested for planning "massive" bombings of discotheques, airports, pubs, and nightspots frequented by Americans in Germany.[17] His death was a huge relief for U.S., Uzbekistani, and German authorities and helped disrupt terrorist attacks against the West.

On October 21, 2009, Abu Musa al Masri, an Egyptian explosives expert and trainer, was killed by a drone in North Waziristan. On December 8 Saleh el Somali, a leader of al Qaeda's foreign operations, was killed while traveling in his car in North Waziristan.[18] On December 17 Zuhaib al Zahib, a commander of the al Qaeda fighting force known as the Lashkar al Zil (Shadow Army), was killed in North Waziristan. During this attack, the CIA took the unusual step of using five drones to fire ten missiles at the compound where Zahib was staying to make sure he did not escape.[19] Sometime after the Zahib strike, Abdullah Said al Libbi, a Libyan who was head of the Lashkar al Zil, was also killed by a drone. Finally, on December 31 an important Taliban commander who was involved in attacks on the Pakistani army and Afghanistan, Haji Omar, was killed, along with his son and a woman, when he was hit by a missile in a *hujra* in North Waziristan.[20]

Also reported killed in 2009 in an unspecified drone strike was Saad bin Laden, Osama's third oldest son.[21] Saad was believed to have been involved in the bombing of a Tunisian synagogue that killed nineteen predominantly German and French tourists in 2002.

By year's end more than sixteen top al Qaeda and Taliban militants and more than five hundred lower-level fighters had been killed. The majority of those who were killed in the strikes were described as "foot soldiers," not HVTs. A U.S. official spoke of this remarkable trend, saying, "This effort has evolved because our intelligence has improved greatly over the years, and we're able to identify not just senior terrorists, but also al-Qaeda foot soldiers who are planning attacks on our homeland and our troops in Afghanistan. . . . We would be remiss if we didn't go after people who have American blood on their hands. To use a military analogy, if you're only going after the generals, you're likely to be run over by tanks."[22]

Hunting for lower-level operatives was not the only trend that began at this time. According to the media sources available for 2009, in total as many as 527 militants and terrorists were killed at a collateral damage cost of forty civilians (and twenty-four unknown). This meant that for the year 2009 roughly 7 percent of those who were killed in the drone campaign were listed by the media as "civilians." The following list is a breakdown of the drone strikes of 2009 with sources from the Pakistani and Western media that led to this startling conclusion:

1. January 1, S. Waziristan, five terrorists killed, no civilians[23]
2. January 2, S. Waziristan, four unknowns killed[24]
3. January 23, N. Waziristan, ten militants (four of them foreigners) killed[25]
4. January 23, S. Waziristan, one Taliban, four civilians killed[26]
5. February 14, S. Waziristan, twenty-five militants killed (among them Arab and Uzbek fighters), no civilians reported killed[27]
6. February 16, Kurram Agency, thirty-one people killed at a "militant hideout," described as a "camp of Afghan Commander Bahram Khan Koch" (presumed militants)[28]
7. March 1, S. Waziristan, eight militants killed, no civilians[29]
8. March 12, Kurram Agency, twenty-two militants killed, no civilians[30]
9. March 15, Bannu, North-West Frontier Province, five militants killed (including two Arabs), no civilians[31]
10. March 25, S. Waziristan, eight "foreign militants" killed in a convoy, no civilians[32]
11. March 26, N. Waziristan, four unknowns killed[33]
12. April 1, Orakzai Agency, fourteen Taliban and al Qaeda members killed, no civilians[34]
13. April 4, N. Waziristan, four to ten militants (including foreigners) killed, seven foreign women and children also reported killed[35]
14. April 8, S. Waziristan, three to four militants killed, no civilians[36]
15. April 19, S. Waziristan, three to seven militants killed, no civilians[37]

16. April 29, S. Waziristan, approximately four Taliban militants and two unknowns killed[38]
17. May 9, S. Waziristan, ten Taliban killed, no civilians[39]
18. May 12, S. Waziristan, eight foreigners and local Taliban killed, no civilians[40]
19. May 16, N. Waziristan, eight local militants and two Arabs killed, no civilians[41]
20. June 14, S. Waziristan, five militants killed, no civilians[42]
21. June 18, S. Waziristan, nine Taliban fighters killed, no civilians[43]
22. June 18, S. Waziristan, five killed at "hideout of Taliban commander Mullah Nazir"[44]
23. June 23, S. Waziristan, six militants killed, no civilians[45]
24. June 23, S. Waziristan, as many as seventy killed at a funeral, including eighteen civilians and approximately fifty-two described as "militants" [46]
25. July 3, S. Waziristan, thirteen militants killed, no civilians[47]
26. July 7, S. Waziristan, sixteen militants killed, no civilians[48]
27. July 8, S. Waziristan, fifty suspected militants killed, no civilians[49]
28. July 8, S. Waziristan, seventeen militants killed, no civilians[50]
29. July 11, S. Waziristan, eight Taliban killed, no civilians[51]
30. July 17, N. Waziristan, five Taliban killed, no civilians[52]
31. August 5, S. Waziristan, one to three Taliban (Baitullah Mehsud and his guards) and one civilian (his wife) killed
32. August 11, S. Waziristan, twelve extremists killed, no civilians[53]
33. August 21, N. Waziristan, twelve unknowns killed[54]
34. August 27, S. Waziristan, eight militants killed, no civilians[55]
35. September 7, N. Waziristan, five fighters killed, no civilians reported killed[56]
36. September 8, N. Waziristan, eight militants killed, no civilians[57]
37. September 14, N. Waziristan, four foreign militants and four unknowns killed[58]
38. September 24, N. Waziristan, twelve Afghan Taliban militants killed, no civilians[59]
39. September 29, S. Waziristan, five suspected Taliban killed, no civilians[60]

40. September 29, N. Waziristan, seven unknowns killed[61]
41. September 29, Khyber Agency, no deaths[62]
42   September 30, N. Waziristan, eight militants killed, no civilians[63]
43. October 15, N. Waziristan, four suspected militants killed, no civilians[64]
44. October 24, Bajaur Agency, twenty-two terrorists killed, no civilians[65]
45. November 4, N. Waziristan, four unknowns killed[66]
46. November 18, N. Waziristan, four militants killed, no civilians[67]
47. November 20, N. Waziristan, eight militants killed, no civilians[68]
48. December 7, N. Waziristan, three militants killed, no civilians[69]
49. December 8, N. Waziristan, three unknowns killed[70]
50. December 17, N. Waziristan, two militants killed, no civilians[71]
51. December 17, N. Waziristan, twelve suspected Taliban militants killed, no civilians[72]
52. December 18, N. Waziristan, eight Taliban killed, no civilians[73]
53. December 26, N. Waziristan, thirteen militants killed, no civilians[74]
54. December 31, N. Waziristan, three militants killed, no civilians[75]

Although it is based on news service reports (usually informed by local Pakistani sources), the preceding list helps explain why Obama, a president who was trying to improve America's frayed, post–Iraq invasion image among Muslims, continued the unpopular drone strikes. Simply put, he had to have known that the CIA was waging perhaps the most precise bombing (or, more accurately, "guided-missile") campaign in world history. Clearly the combination of high-resolution optics and spies on the ground was working to minimize civilian casualties, even as hundreds of militants and terrorists were killed with surgical precision. Many of those who were killed were involved in mass-casualty terrorism plots that would have taken the lives of many more civilians had the plotters not been killed. Certainly, Obama and the CIA felt they were saving civilian lives by killing those who murdered innocents in Pakistan, Afghanistan, Uzbekistan, Africa, and the West. The bad public relations fallout in Pakistan and the blowback of hatred for America among families of slain militants and the few slain civilians was apparently worth it.

Thus, the Obama administration made several attempts to change the widely held perception that the drones were killing predominantly civilians. One U.S. official, for example, described the drones' much vaunted accuracy in the following exaggerated terms:

In the past year, in the neighborhood of 600 militants—including more than two dozen terrorist leaders—have been taken off the battlefield. In the same period we can't confirm any noncombatant casualties. This is a weapon—fuelled by good intelligence—that allows us to counter an urgent and deadly threat in otherwise inaccessible places. And it's far more precise than conventional ground operations. What's the alternative to this kind of rigor, assuming the United States and its allies are unwilling to allow al-Qaeda and its friends to plot and murder freely?[76]

Another U.S. official similarly claimed that the CIA had killed two thousand militants and just fifty noncombatants in the period of 2001–2011 (i.e., a ratio of 2.5 percent civilian deaths).[77] On a separate occasion Deputy Homeland Security Adviser John Brennan emphatically stated,

One of the things President Obama has insisted on is that we're exceptionally precise and surgical in terms of addressing the terrorist threat. And by that I mean, if there are terrorists who are within an area where there are women and children or others, you know, we do not take such action that might put those innocent men, women and children in danger. In fact I can say that the types of operations that the US has been involved in, in the counter-terrorism realm, that nearly for the past year there hasn't been a single collateral death because of the exceptional proficiency, precision of the capabilities that we've been able to develop.[78]

The preceding casualty list indicates that the basic thrust of Brennan's statement—namely, that the CIA was making every effort to avoid civilian casualties in the FATA—was true, although the deputy adviser was obviously exaggerating when he said there hadn't been a "single collateral death." For his part, President Obama said, "I want to make sure that people understand that actually, drones have not caused a huge number of civilian casualties. For the most part, they have been very precise, precision strikes against al-Qaida and their affiliates. And we are very careful in terms of how it has been applied."[79] Obama rejected the notion that "we're just sending in a whole bunch of strikes willy nilly." Stressing the pinpoint nature of the strikes, he said, "This is a targeted, focused effort at people who are on a list of active terrorists, who are trying to go in and harm Americans, hit American facilities, American bases."[80]

In December 2009 the *New York Times* published an article based on an interview with a U.S. government official that stated,

> Assessments of the drone campaign have relied largely on sketchy reports in the Pakistani press, and some have estimated several hundred civilian casualties. Saying that such numbers are wrong, one government official agreed to speak about the program on the condition of anonymity. About 80 missile attacks from drones in less than two years have killed "more than 400" enemy fighters, the official said, offering a number lower than most estimates but in the same range. His account of collateral damage, however, was strikingly lower than many unofficial counts: "We believe the number of civilian casualties is just over 20, and those were people who were either at the side of major terrorists or were at facilities used by terrorists."[81]

The CIA itself finally entered the fray when one of its spokesmen made the following statement: "While the C.I.A. does not comment on reports of Predator operations, the tools we use in the fight against Al Qaeda and its violent allies are exceptionally accurate, precise and effective. Press reports suggesting that hundreds of Pakistani civilians have somehow been killed as a result of alleged or supposed U.S. activities are—to state what should be obvious under any circumstances—flat-out false."[82]

A 2011 article published for Bloomberg supported the CIA statement and reported that the number of civilian deaths resulting from drone strikes had plummeted. According to this article, titled "U.S. Said to Reduce Civilian Deaths after Increasing CIA Pakistan Strikes,"

The Central Intelligence Agency, while increasing the frequency of drone strikes in Pakistan, has reduced civilian casualties, a U.S. official and independent analysts said. The 75 strikes launched in the ungoverned tribal region since the drone program accelerated in mid-August have killed several hundred militants without causing any deaths among civilian non-combatants, said the U.S. official, who, lacking authorization to discuss the program, requested anonymity. Analysts who monitor developments in the region said figures based on press reports show a decline in unintended deaths, although verifying exact figures may be impossible. "The drone strikes do appear to be becoming less lethal for civilians as time goes on," said Bill Roggio, editor of *The Long War Journal*.[83]

As this article states, some independent scholars, such as Roggio, backed up the government's claims to small numbers of civilian casualties. Georgetown professor Christine Fair even went so far as to declare on television, "Actually the drones are not killing innocent civilians. Many of those reports are coming from deeply unreliable and dubious Pakistani press reports, which no one takes credibly on any other issue except for some reason on this issue. There've actually been a number of surveys on the ground, in FATA. The residents of FATA generally welcome the drone strikes because they know actually who's being killed. They're very much aware of who's being killed and who's not."[84]

Like John Brennan and the unnamed government official quoted in the Bloomberg article, Fair exaggerated to make her point. No one outside of the government could realistically argue that the drones didn't kill any civilians; my own study shows forty civilians were killed in the year 2009 alone. But the greater point—that the drones were not clumsily killing mass numbers of civilians—is supported by the data in the preceding casualty list.

The media reports on which the casualty list is based are the products of journalists' interviews with civilians, Taliban, and local officials from the targeted areas. But the journalists, even the Pakistani reporters, rarely gain access to the actual targeted strike zones. The Taliban distrust even Pakistani journalists who are generally opposed to the strikes on their country. Therefore, scholars trying to assess the number of slain civilians in the strikes perforce have to rely on secondary media accounts from Western and Pakistani journalists who have access to local civilian, government, and Taliban sources. (The Taliban seem to be remarkably forthcoming in discussing the deaths of their comrades.)

One widely quoted study based on the available media reports demonstrates that the drones kill predominantly noncivilians. Peter Bergen and Katherine Tiedemann at the New America Foundation carried out a case-by-case, sourced study of drone strike fatalities that concluded,

> We've been able to discern some surprising trends. A frequent criticism of the drones program is that the strikes kill too many civilians. . . . But even as the number of reported strikes has skyrocketed—with one every three days in 2010, compared with one a week last year and one every 11 days in 2008—the percentage of non-militants killed by the attacks has plummeted. . . . If the true nonmilitant fatality rate were more widely known in Pakistan, the program might be less unpopular there. Those targeted in the strikes, after all, are thought to have carried out or planned attacks not only in Afghanistan and the West, but also in Pakistan, where more than 4,000 people have been killed in militant attacks since the Red Mosque incident in July 2007.[85]

These authors further wrote, "Our study shows that the 265 reported drone strikes in northwest Pakistan, including 52 in 2011, from 2004 to the present have killed approximately between 1,628 and 2,561 individuals, of whom around 1,335 to 2,090 were described as militants in reliable press accounts. Thus, the true non-militant fatality rate since 2004 according to our analysis is approximately 20 percent. In 2010, it was more like five percent."[86] By 2012 the civilian death rate had fallen even further, and Bergen boldly wrote, "Today, for the first time, the estimated civilian death rate is at or close to zero."[87]

This independent study's conclusion that the drones were killing civilians at a rate of roughly 5 percent in 2010 seems to speak to the drones' unprecedented precision. The obvious question, if such findings are accurate, is, How are the drones so precise at killing terrorists and militants? Part of their success can of course be explained by the drone pilots' ability, while flying remotely from Langley, Creech Airbase in Nevada, or elsewhere, to follow their targets up close for many hours using high-resolution cameras. For the first time in history, unseen drones can patiently wait for targets to gather in a compound, automobile, or training camp; watch to see if civilians are in the area; and then fire when the moment is most opportune. One account compared the drones to "mini-satellites that can monitor a suspected terrorist compound for weeks, watching where the people go and with whom they interact, to help confirm that the right people are being singled out for attack."[88] A *New York Times* article on the drones similarly reported,

> Former C.I.A. officials say there is a rigorous protocol for identifying militants, using video from the Predators, intercepted cell phone calls and tips from Pakistani intelligence, often originating with militants' resentful neighbors. Operators at C.I.A. headquarters can use the drones' video feed to study a militant's identity and follow fighters to training areas or weapons caches, officials say. Targeters often can see where wives and children are located in a compound or wait until fighters drive away from a house or village before they are hit.[89]

According to the *Los Angeles Times*, in some cases drones conduct surveillance for "days" to gather the "evidence that justifies firing a missile." The report went on to say, "One former official directly involved in the program said many locations were watched so closely that the CIA could predict daily routines. 'Is the white van there yet?' the official said, giving an example of the degree of scrutiny. 'Is he walking with a limp?'"[90]

There are additional safeguards in place as well. First, the CIA's Covert Action Review Group leads a debate on the "kill"; then, the target is passed to the CIA's Counterterrorism Center. The CIA drones cannot fire unless their pilots receive final approval from the CIA director or his deputy.[91] Describing former CIA chief Leon Panetta's direct involvement in the strikes, one senior intelligence official said, "He asks a lot of questions about the target, the intelligence picture, potential collateral damage, women and children in the vicinity." For his part, Panetta claimed this oversight meant that the drones were "very precise" and "very limited in terms of collateral damage."[92] Panetta also said, "You know, as a Catholic, I remember when I first became director of the C.I.A., and realized that I was making life-and-death decisions—with regards to our operations. It doesn't come lightly. . . . Frankly, we made very clear that if there were any women or children that were involved that we would not take a shot. That became a rule that we abided by."[93]

In a rare discussion of the topic, President Obama similarly said, "We are very careful in terms of how it's applied. It is very important for everybody to understand that it's kept on a tight leash."[94] Obama further stated in September 2012, "We will not engage in operations if we think there's going to be civilian casualties involved."[95] The *Los Angeles Times* reported that House and Senate intelligence committee members monitor the strikes and are informed about them. One Democrat member of the committee said of the process, "If the American people were sitting in the room, they would feel comfortable that it was being done in a responsible way."[96] Obama's chief counterterrorism adviser, John Brennan, who has played a key and often restraining role in the drone strikes, stressed during his February 2013 hearing testimony that the drones are used only "as a last resort to save lives when there is no alternative."[97] Although Brennan has been dubbed the "assassination czar" by his critics for his involvement in the drone campaign, insiders at the CIA consider him to be "a rein, a constrainer. He is using his intimate knowledge of intelligence and the process to pick apart their arguments that might be expansionary."[98]

Other reports have said that the U.S. ambassador to Pakistan must approve drone hits to make sure there is no political fallout from a strike and that this has even led to clashes between the CIA station chief in Islamabad and the ambassador.[99] In order to carry out a strike, there must be two forms of supporting intelligence, for example, a radio intercept, a report by a spy, or a visual from a drone. If these steps were not enough, CIA lawyers must also sign off on all HVTs who are on a hit list. *Newsweek* described a typical precision drone attack on an enemy HVT that was approved by CIA lawyer John Rizzo:

> It was an ordinary-looking room located in an office building in northern Virginia. The place was filled with computer monitors, keyboards,

and maps. Someone sat at a desk with his hand on a joystick. John A. Rizzo, who was serving as the CIA's acting general counsel, hovered nearby, along with other people from the agency. Together they watched images on a screen that showed a man and his family traveling down a road thousands of miles away. The vehicle slowed down, and the man climbed out.

A moment later, an explosion filled the screen, and the man was dead. "It was very businesslike," says Rizzo. An aerial drone had killed the man, a high-level terrorism suspect, after he had gotten out of the vehicle, while members of his family were spared. "The agency was very punctilious about this," Rizzo says. "They tried to minimize collateral damage, especially women and children."[100]

Another official described an incident wherein a drone pilot was able to divert a guided missile from an intended target at the last minute to save civilians: "In one recent strike, an official said, after the drone operator fired a missile at militants in a car and a noncombatant suddenly appeared nearby, the operator was able to divert the missile harmlessly into open territory, hitting the car minutes later when the civilian was gone."[101]

An *Esquire* magazine journalist who was embedded with a drone crew in Afghanistan similarly reported,

> In preparation for raids or missile strikes, crews sometimes loiter over an area for weeks, building video dossiers. . . .
>
> If they're tracking an individual, as they often will for days or weeks, they know when he goes to work, where he stops for tea, and whom he talks to along the way. Though civilians do die in some of the missile strikes, this ability to linger can do much to limit unintended deaths. If women and children or the unlucky neighbor is nearby, the plane can wait, and wait, without losing sight.[102]

Such safeguards have prevented civilian deaths from being higher, according to the *Los Angeles Times*. In an article titled "CIA May Be Avoiding Civilian Deaths," the *Los Angeles Times* reported that the CIA had passed up a chance to kill Sirajuddin Haqqani, the son and heir to the notorious

Afghan Taliban terrorist Jalaluddin Haqqani, in 2010 because there was an unacceptably large number of civilian women and children near him when a drone spotted him.[103]

But bureaucratic safeguards and unprecedented ability to covertly monitor potential targets from afar for hours at a time using high-resolution optics are not the only explanation for the low civilian death toll. In September 2010 the *Washington Post* reported that the CIA had begun to run a program from bases in the border region of Pakistan from which they controlled spies in the FATA. The CIA had set up counterterrorist pursuit teams made up of Afghan and Pakistani Pashtuns who ranged beyond the borders of Afghanistan into the FATA, hunting Taliban and al Qaeda militants.[104] These spies reported the whereabouts of the militants to CIA drones, which then took them out.

A Pakistani journalist quoted at the time had some interesting insights into how the CIA had begun to choose its targets with the use of local spies and surveillance: "They [the locals] see traffic coming and going from the fortress homes of tribal leaders associated with foreign elements, and they pass the information along. Some quick surveillance is done, and then someone pops a couple of hundred-pound bombs at the house."[105]

In his book *Obama's Wars*, Bob Woodward wrote of the importance of these spies in his discussion of the transfer of power from Bush to Obama in 2008–2009:

> But, he [President Bush] said, the real breakthrough had been with human sources. That is what President Bush wanted to protect at all costs. The drones were basically flying high-resolution video cameras armed with missiles. The only meaningful way to point a drone towards a target was to have spies on the ground telling the CIA where to look, hunt and kill. Without spies, the video feed from the Predator might as well be a blank television screen.
>
> McConnell provided extensive details about these human sources who had been developed in an expensive, high-risk program over five years. The spies were the real secrets that Obama would carry with him from that moment forward. They were the key, in some respects, to protecting the country.[106]

The Pakistani ISI and military also stepped up their cooperation with the CIA in providing information on targets. One Pakistani official claimed that "as a result the strikes are more and more precise."[107] To enhance cooperation with the Pakistanis who were clearly acting as spotters, the United States set up a coordination center with them at Torkham Gate, on the Afghan-Pakistani border near the Khyber Agency, and there they shared footage from drone strikes with their Pakistani counterparts.[108]

The Taliban quickly came to conclusion that the drones' unique ability to target their fighters was the result of the CIA's reliance on a network of local spies. One Taliban spokesman said, "The growing drone attacks in North Waziristan are proving that someone on the ground is guiding the spy planes to strike targets. The improvement in hitting accurate targets, as witnessed in the recent weeks, is nothing but an indication that the CIA-paid agents are very much active."[109]

Spying activities were also noticed by al Qaeda, which published an online book titled *Guide to the Laws Regarding Muslim Spies*. In it, al Qaeda leader Abu Yahya al Libi stressed the danger of spies to his organization: "It would be no exaggeration to say that the first line in the raging Crusader campaign waged by America and its allies against the Muslims and their lands is the network of spies."[110] He went on to lament the spies' negative impact on al Qaeda, writing, "One single piece of information transmitted to them [the Americans], by one of their spies, is able to exasperate spirits, honor, and possessions in a way that thousands of their mobilized soldiers cannot do."[111] Libi then writes in paranoid terms about the seemingly ubiquitous presence of the spies in the FATA:

> The spying networks are their eyes to see the hidden things that they cannot see and are their hands that are still extending inside the houses, in the forests, up the mountains, into the valleys, and inside the dark caves in order to catch a target that their developed technology was not able to reach. The spies are the brigades, the soldiers, and they are present and absent at the same time. They were sent to penetrate the ranks of the Muslims generally, and the mujahidin specifically, and spread all over the lands like locusts.
>
> Although the spies are busy day and night carrying out their duties in an organized and secret manner and taking directions, even orders like

soldiers, still you never feel their presence. You can see their influence like killing, destroying, imprisoning, and tracking, but you do not see them.[112]

In May 2009 the British newspaper the *Guardian* published a story that revealed for the first time a fourth means by which the CIA drones had been able to track their targets with uncanny precision. The *Guardian* reported that the CIA spies in the FATA region had been planting small homing beacons known as *pathrai* (chips) in the vehicles, compounds, *hujras*, and camps of al Qaeda and Taliban militants. These infrared flashing beacons were previously used by the military to identify friend from foe, mark drop zones, and outline perimeters.[113] Now the drones picked up the signals emitted by the beacons and were able to fire their missiles precisely at the targets beings designated, or "lit up," by the *pathrai*. According to the *Guardian*,

> "Everyone is talking about it," said Taj Muhammad Wazir, a student from south Waziristan. "People are scared that if a pathrai comes into your house, a drone will attack it." According to residents and Taliban propaganda, the CIA pays tribesmen to plant the electronic devices near farmhouses sheltering al-Qaida and Taliban commanders. Hours or days later, a drone, guided by the signal from the chip, destroys the building with a salvo of missiles. "There are body parts everywhere," said Wazir, who witnessed the aftermath of a strike. . . . For the US military, drones have proved to be an effective weapon against al-Qaida targets, and they are becoming increasingly accurate.[114]

The targets of the attacks quickly understood that the drones' precision was no coincidence. The Taliban and al Qaeda, as previously mentioned, became vitally aware that tribesmen around them, and even some of their own members, had been hired to spy on them and relay their location to the CIA drones through various means, including *pathrai* homing beacons. In al Qaeda's Internet book, the organization wrote of the chips, "These result in the firing of the murderous and destructive missiles whose wrath is inflicted on the Mujahedeen and the weak." The book includes photos of some of the devices that "spies painstakingly transport to the targets they are assigned by their infidel patrons."[115]

The Taliban were shocked by the accuracy of the strikes guided by the *pathrais*. One *Newsweek* account of a drone strike on a Taliban safe house that killed several al Qaeda operatives and their Afghan allies reported,

> After one of the latest U.S. Predator attacks in North Waziristan, a Taliban subcommander visited the site. He's seen the results of many airstrikes over the past year or two, but this one really impressed him. The missile didn't just hit the right house; it scored a direct hit on the very room where Mustafa al-Misri ("Mustafa the Egyptian") and several other Qaeda operatives were holed up. The hit was so accurate, the subcommander says, it's as if someone had tossed a GPS [global positioning system] device against the wall. Unfortunately for others at the scene, the mud-and-stone house collapsed, killing several Afghans along with the foreign fighters. Nevertheless, the subcommander told Newsweek, "We are stunned" by such precision.[116]

Al Qaeda has similarly bemoaned the existence of "spies who have spread throughout the land like locusts."[117]

To combat these spies, according to local sources, the Taliban launched a witch hunt in the FATA, which has resulted in the deaths of real or suspected spies for giving away militants' positions to the CIA.[118] The beheaded or mutilated bodies of locals who were said to have been spying for the Americans or "Indian and Jewish agencies" have been turning up in record numbers since the drone campaign began.[119] According to the *Times* of London, "After every raid witnesses say that the Taleban react with rage, abducting, torturing and killing anyone suspected of planting a chip. 'Sometimes we see a body a day lying by the roadside,' said Gul Rafay Jan, from Miran Shah. 'They've got signs around their necks saying they were spies planting chips. Sometimes they have been tortured to make confession videos by having rods pushed through their arms or stomachs, or being suspended over a fire.'"[120]

As the CIA began to focus on hitting automobiles carrying Taliban and al Qaeda members, the militants even began executing dozens of car mechanics whom they blamed for placing *pathrais* in their trucks and cars. So concerned were the Pakistani Taliban about local spies that they created a special force in North Waziristan, known as the Lashkar al Khorasan (the Army of Greater Afghanistan) or the Khorasan Mujahideen, to hunt down and execute real or perceived spies.[121] This two-hundred-man task force of masked killers whose faces are covered in black balaclavas began kidnapping and killing real or suspected spies on an almost "daily basis" and terrorized much of North Waziristan. One Pakistani source said, "According to local tribal elders, in most cases militants execute so-called spies just to terrorize ordinary tribesmen. In some cases, the militants are also known to put suicide vests on those accused of spying and detonate the vests in front of large crowds to demonstrate the power of the Taliban."[122] In another case, "the bodies [of suspected spies] were beyond recognition because militants had repeatedly shot the victims in the face."[123]

American reporter David Rohde, who was captured by the Haqqani Network and held prisoner in North Waziristan, saw for himself the Taliban's efforts to hunt down real or suspected spies. His extraordinary eyewitness account reveals some insight into their feverish witch hunts:

> After 15 minutes, the guards took me back to our house and explained what had happened. Missiles from American drones had struck two cars, they said, killing seven Arab militants and local Taliban fighters. Later, I learned that one of our guards suggested I be taken to the site of the attack and ritually beheaded. The chief guard overruled him.
>
> The strikes fueled a vicious paranoia among the Taliban. For months, our guards told us of civilians being rounded up, accused of working as American spies and hung in local markets. Immediately after that attack in South Waziristan, a feverish hunt began for a local spy who the Taliban were convinced had somehow secretly guided the Americans to the two cars.
>
> Several days after the strike, our guards told us foreign militants had arrested a local man and accused him of guiding the drones. After the jihadists disemboweled the villager and chopped off his leg, he "confessed" to being an American spy, they said. Then the militants decapitated the man and hung his corpse in the local bazaar as a warning.[124]

Clearly these cruel actions reveal that the furious Taliban and al Qaeda fighters understood that the CIA was targeting them exclusively—not random civilians.

The *pathrais* were not the final piece of the puzzle of the drones' uncanny precision. The missiles the drones fired had also become smaller and less likely to kill innocents via large explosions or shrapnel. In April 2010 the *Washington Post* published a groundbreaking article that reported,

> The CIA is using new, smaller missiles and advanced surveillance techniques to minimize civilian casualties in its targeted killings of suspected insurgents in Pakistan's tribal areas, according to current and former officials in the United States and Pakistan. The technological improvements have resulted in more accurate operations that have provoked relatively little public outrage, the officials said. . . .
>
> Last month, a small CIA missile, probably no bigger than a violin case and weighing about 35 pounds, tore through the second floor of a house in Miran Shah, a town in the tribal province of South Waziristan. The projectile exploded, killing a top al-Qaeda official and about nine other suspected terrorists. The mud-brick house collapsed and the roof of a neighboring house was damaged, but no one else in the town of 5,000 was hurt, according to U.S. officials who have reviewed after-action reports. . . .
>
> The clamor over the strikes has died down considerably over the past year and Pakistani officials acknowledge that improved accuracy is one of the reasons. Pakistani security officials say that better targeting technology, a deeper pool of spies in the tribal areas, and greater cooperation between the U.S. and Pakistani intelligence services have all led to strikes that cause fewer civilian deaths.[125]

The violin case–sized missile described in this article appears to be the thirty-five-pound Lockheed Martin Scorpion. On its website Lockheed described the Scorpion as the "war fighter's answer to precision attack using a small, lethal warhead against targets in areas requiring low collateral damage." The site further declared, "The precision provided by these seeker types ensures accuracy to less than one meter and dramatically reduces the possibility of collateral damage."[126] Whereas the Hellfire missile previously fired by drones was sixty-four inches long and weighed 108 pounds, the Scorpion was twenty-one inches in length and weighed thirty-five pounds. Although the drones could easily shoot a salvo of Hellfires or safely drop heavier Paveway laser-guided bombs on training camps, the more surgical approach required in "civilian-rich" environments, where innocent bystanders could be accidentally killed, meant the Scorpion. The new smaller missiles allowed the United States to proudly proclaim that although in Vietnam the collateral damage radius for an aerial bomb explosion was four hundred feet, with a drone it had been diminished to forty feet.[127]

The *Economist* pointed to yet a sixth reason for the plummeting number of civilian deaths from drone strikes, namely, the tendency for the CIA to hit targets that were in vehicles, "where militants can be more easily hit without killing civilians."[128] In essence, the drones could see who got into the vehicles and then target them when they were driving in areas with few civilians nearby. An extensive report by the Indo-Asian News Service (IANS) found that

As many as 39 precision strikes targeting Al Qaeda and its Pakistani affiliates in North Waziristan took place over the past three months, an [American] official said, adding the strikes killed 605 people. The dead comprised 507 Pakistanis, majority of them militants, and 98 foreigners.

Security officials have been taken by surprise by the CIA's increasing ability to take out moving targets. Predator drones Monday fired missiles at two vehicles in North Waziristan, killing 18 people.

"One of the vehicles was loaded with explosives to the hilt and had it been targeted in a compound the devastation would have been huge," the official was quoted as saying. "So a moving target is ideal in the sense that it minimizes chances of collateral damage."

Officials said there has been mounting evidence of the CIA tracking moving targets in Pakistan's tribal regions from inside Afghanistan and then attacking them. "The evidence we have is circumstantial but that the CIA is able to hit mobile target demonstrates enhanced humint on the ground," the official said on the condition of anonymity.

A source said: "The Americans seem to have made considerable ingress in our tribal regions and I doubt this could have happened without our knowledge and approval." The security official pointed out that "real-time [human] intelligence" in North Waziristan had helped the CIA to hit moving targets.

"They have improved their intelligence collection to deliver punishment in real time," said another official while referring to the drone attacks in the tribal areas. "Moving targets tend to vanish quickly. So you have to have human intelligence on the ground to identify and engage the target in real time in a matter of minutes. This requires credible intelligence and communication system to direct the strike and this means that CIA's human intelligence has improved considerably," the official said.[129]

At this time there were many reports in the Pakistani press about Taliban and al Qaeda vehicles being destroyed by drones, presumably based on a combination of on-the-ground humint and drone surveillance. A typical account read,

In another attack by the US spy planes, a double-cabin pickup truck was targeted in Mezar village of Dattakhel Tehsil not far from the Urgoon area of Afghanistan's Paktika province. Official sources said six militants, suspected to be foreigners, were killed in the attack. "The pickup truck was split into pieces and there was almost no sign of the five people travelling in the vehicle," Taliban sources told our sources. The sources said those killed were Arab fighters returning to their hideouts located in the mountainous border areas between Pakistan and Afghanistan after a clash with the US-led coalition forces across the border in Paktika. However, details about their nationality and identity weren't available.[130]

New York Times correspondent David Rohde similarly commented on the drones' accuracy in hitting Taliban vehicles. He wrote, "Based on the reactions of the [Taliban] guards, the attacks appear to primarily kill militants." Among other things, Rohde described being near a drone attack on two cars that killed seven Arab militants and a local Taliban fighter, but no civilians.[131]

In March 2011 the *New York Times* reported, "In recent years they [the drone strikes] have provoked less outrage in the tribal areas, as the strikes have focused increasingly on foreign fighters loyal to Al Qaeda who have infiltrated the area, and as fewer civilians have been killed by them."[132] Christine Fair, who carried out fieldwork in the FATA, reported, "When children hear the buzz of the drones, they go to their roofs to watch the spectacle of precision rather than cowering in fear of random 'death from above.'"[133]

Pakistani military officials on the ground in the region, who had intimate knowledge of who was being killed and who was not, also remarked on the drones' increasing accuracy. One senior Pakistani official said, "You don't hear so much about it [the drone campaign]. . . . There are better targets and better intelligence on the ground. It's less of a crapshoot."[134] According to the Wikileak documents that exposed many aspects of the Pakistani and U.S. views on drones,

> A U.S. diplomat, based in Peshawar near the border territories, mentions in a 2008 cable a meeting he had with a senior [Pakistani] official whose name is redacted. The official "said he wanted to say in an unofficial capacity that he and many others could accept Predator strikes as they were surgical and clearly hitting high value targets. He mentioned that fear among the local populace in areas where the strikes have been occurring was lessening because 'everyone knew that they only hit the house or location of very bad people.'"[135]

Similarly, Malik Naveed Khan, a provincial police chief from the tribal zone, could not contain his admiration for the drones: "They are very precise, very effective, and the Taliban and al-Qaida dread them."[136] A former Pakistani ambassador further supported the claims of accuracy, stating, "There is no gainsaying the fact that drone accuracy has improved considerably. Of late only those get killed who are targeted. And, by the looks of it, there is no more effective way of reaching the upper echelons of the Taliban."[137] A former Pakistani general opined, "There is no doubt that the

US military and the CIA, with better intelligence and sophisticated technology, have been more accurate in their targeting. They have also been successful in hitting at least mid-level Al Qaeda and Taliban leaders, and with less collateral damage. Apparently, improved targeting has been possible by integrating the latest technology with a reliable network of human intelligence that places transmitters at the right places for the drones to respond."[138]

Pakistani citizens also noticed the change. A Pakistani blogger on the popular Khyber Watch site wrote, "According to our latest and fresh information obtained from friends and political people that whenever drones are seen hovering in Waziristan our people are satisfied that only terrorists and their friends will be hit hard."[139] Similarly, Professor Ijaz Khattak of the University of Peshawar explained to a popular television host in Pakistan, "The drone attacks have proved effective and have targeted the terrorists and there has been little collateral damage in the US drone attacks."[140] A Pashtun researcher from the region interviewed Pashtun students from the FATA and found that these students thought that "the drone attacks cause a minimum loss of innocent civilians and their property. The respondents appreciated the precision of such attacks."[141] A resident of North Waziristan who witnessed numerous missile strikes told the American Broadcasting Company (ABC), "The attacks have become so precise. In a village, if they want to hit a house in the middle of the village and it's surrounded by other houses, the missile would come and hit that one house only."[142]

Surprisingly, the Taliban acknowledged this precision as well. One pamphlet issued by the Taliban in North Waziristan acknowledged, "Westerners have some regard for civilians, and they do distinguish between Taliban fighters and civilians, but the Pakistani army doesn't."[143] Articles that reported the drones' precision in killing al Qaeda and Taliban targets and avoiding civilian deaths also occasionally appeared in the Pakistani media. A December 2010 article in the Pakistani newspaper *Dawn*, for example, read, "American officials do not acknowledge that war or discuss who is being killed in drone-fired missile attacks on al-Qaida and Taliban targets, which have surged this year to average about two a week. But they have said privately that the strikes are highly precise and harm very few innocents. Some locals agree about their accuracy, especially when compared to bombing runs by Pakistani jets."[144] A Pashtun from the FATA agreed with this when she wrote of her experience in at home:

> I have heard people particularly appreciating the precision of drone strikes. People say that when a drone would hover over the skies, they wouldn't be disturbed and would carry on their usual business because they would be sure that it does not target the civilians, but the same people would run for shelter when a Pakistani jet would appear in the skies because of its indiscriminate firing. They say that even in the same compound only the exact room [where an HVT is present] is targeted. Thus others in the same compound are spared.[145]

But the most important testimony to the drones' accuracy was yet to come. On March 9, 2011, *Dawn* published a remarkably frank interview with a member of the Pakistani military based in Waziristan who openly supported the drone strikes on the basis of their precision in killing terrorists. The interview, titled "Most of Those Killed in Drone Attacks Were Terrorists," was electrifying for the minority of Pakistanis who supported the drone strikes and was scorned by those who detested them. The article read,

> In a rather rare move, the Pakistan military for the first time gave the official version of US drone attacks in the tribal region and said that most of those killed were hardcore Al Qaeda and Taliban terrorists and a fairly large number of them were of foreign origin.
>
> General Officer Commanding 7-Division Maj-Gen Ghayur Mehmood said in a briefing here: "Myths and rumors about predator strikes and the casualty figures are many, but it's a reality that many of those being killed in these strikes are hardcore elements, a sizeable number of them foreigners. Yes there are a few civilian casualties in such precision strikes, but a majority of those eliminated are terrorists, including foreign terrorist elements."
>
> The Military's 7-Division's official paper on the attacks till Monday said that between 2007 and 2011 about 164 predator strikes had been carried out and over 964 terrorists had been killed. Of those killed, 793

were locals and 171 foreigners, including Arabs, Uzbeks, Tajiks, Chechens, Filipinos and Moroccans.

In 2007, one missile strike left one militant dead while the year 2010 was the deadliest when the attacks had left more than 423 terrorists dead. In 2008, 23 drone strikes killed 152 militants, 12 of them were foreigners or affiliated with Al Qaeda.

In 2009, around 20 predator strikes were carried out, killing 179 militants, including 20 foreigners, and in the following year 423 militants, including 133 foreigners, were killed in 103 strikes. In attacks till March 7 this year, 39 militants, including five foreigners, were killed.

Maj-Gen Ghayur, who is in-charge of troops in North Waziristan, admitted that the drone attacks had negative fallout, scaring the local population and causing their migration to other places. Gen Ghayur said the drone attacks also had social and political repercussions and law-enforcement agencies often felt the heat.[146]

A subsequent article in *Dawn* also applauded the frank tone set by General Mehmood: "Is the army hinting that the strikes are a useful and precise tactic in neutralizing identified militants and terrorists? If that is the case, then the military and political leaders should publicly change their stated position and matters should move on—the battle against local and foreign terrorists hiding in the country's north-western regions is far from over."[147]

But such opinions were clearly the minority in Pakistan, and the general's surprising frankness has done little to sway his compatriots' ill will toward the drone strikes. A 2009 Gallup poll found that 67 percent of Pakistanis were opposed to the drone strikes, while 24 percent had neutral feelings toward them and just 9 percent favored them.[148] To a large extent, even those who opposed the Taliban were against the drone strikes for the simple reason that they believed that they were uniquely targeting innocent civilians, not militants. Such contrafactual perceptions have been driven by the Pakistani media, which, as stated earlier, is dominated by journalists who equate drone attacks with random acts of Taliban terrorism.

Several journalists in the Pakistani media have written articles on the drone strikes that have inflated the number of dead civilians and, in so doing, inflamed public opinion against the drones. In April 2009 the Pakistani newspaper the *News*, for example, published an article that completely inverted the low-civilian-casualty trend identified in the preceding case-by-case study of media reports as well as in the New America Foundation report. According to this article, which referenced "cross-border raids" (as previously noted, the drones were actually based in Pakistan), the drones proved to be, in the history of bombing campaigns, uniquely incapable of killing their designated targets. The *News* report stated, without citing any study to back up its claim, "Of the 60 cross-border predator strikes carried out by the Afghanistan-based American drones in Pakistan between January 14, 2006 and April 8, 2009, only 10 were able to hit their actual targets, killing 14 wanted al-Qaeda leaders, besides perishing [*sic*] 687 innocent Pakistani civilians. The success percentage of the US predator strikes thus comes to not more than six per cent."[149] The newspaper reported that this translated to more than fifty civilians killed for every slain al Qaeda member.

Another Pakistani newspaper, *Dawn*, raised the ante and claimed that "of the 44 predator strikes carried out by US drones in the tribal areas of Pakistan over the past 12 months, only five were able to hit their actual targets, killing five key Al-Qaeda and Taliban leaders, but at the cost of over 700 innocent civilians. . . . For each Al Qaeda and Taliban terrorist killed by the American drones, 140 civilian Pakistanis also had to die."[150] This stunning report of course led Pakistanis to believe that the high-tech drones were the most uniquely inaccurate "bombers" in history. The remarkable statistics of fifty or 140 civilians per al Qaeda and Taliban death reported by *Dawn* and the *News* were not, however, backed by any published databases and were actually contradicted by the day-to-day reports of Taliban and al Qaeda deaths found in both newspapers. In fact, a casual perusal of articles on drone strikes in both these newspapers reveals a striking contradiction. In the vast majority of specific articles about drone strikes, reporters described the majority of victims as "militants" or "terrorists," not as "civilians."

A case-by-case analysis of Pakistani and Western reports of drone strikes by me and two of my colleagues at the University of Massachusetts–Dartmouth, Avery Plaw and Matthew Fricker, for the Washington, D.C.–based Jamestown Foundation found that a mere 5 percent of drone-strike victims were described in the media as "civilians."[151] The previously mentioned New America Foundation study similarly found that in 2010 approximately 6 percent of those killed in drone strikes were identified as "civilians."[152] Research conducted by the *Long War Journal* on drone strikes from 2004 to 2011 indicates that approximately 108 civilians were killed in drone strikes that successfully targeted 1,816 Taliban and al Qaeda extremists (i.e., a civilian death rate of less than 6 percent).[153]

These three independent research–based studies, however, failed to have the same impact in Pakistan as the more alarmist, nonsourced findings of the two Pakistani journalists who wrote the aforementioned articles. Afghanistan–Pakistan–based journalist Dexter Filkins of the *New York Times* reported, "The overall perception is that America is massacring people."[154] The exaggerated statistic of fourteen dead al Qaeda for seven hundred civilians was picked up by Pakistani politicians on the right who used these numbers to galvanize popular support against the United States and the drones. One Pakistani writer complained of this trend: "From Imran Khan to Munawar Hasan, right-wing political parties and religious groups have used drone strikes to forward their agenda by misguiding people through erroneous, fabricated and fictional data. As a result, thousands of people have been mobilised across the country to oppose these strikes."[155] Another Pakistani from the FATA similarly complained of this trend among right-wing Pakistanis: "I would request them to stop throwing around fabricated figures of 'civilian casualties' that confuse people around the world and provide propaganda material to the pro-Taliban and al Qaeda forces in the politics and media of Pakistan."[156] A third Pakistani wrote,

> Civilian deaths are not as high as Pakistani media, religious leaders, politicians, and other analysts have been claiming. The analysts question the claims of high civilian casualties because no media outlet or organization has ever published the names of those killed, their villages, dates, and the locations of the drone attacks. According to analysts, in a bid to minimize their losses, the insurgents try to conceal the identities of their associates killed in the attacks. They collect their comrades' bodies and, after burying them, issue statements that all of the victims were innocent residents.[157]

Even though some Pakistanis had challenged the inflated, sourceless Pakistani media "statistics" on civilian deaths, the false numbers were subsequently used to galvanize opinion against the drones in America as well. In

May 2009 David Kilcullen and Andrew Exum published an opinion piece in the *New York Times* that legitimized the sourceless, exaggerated Pakistani media claims: "Press reports suggest that over the last three years drone strikes have killed about 14 terrorist leaders. But, according to Pakistani sources, they have also killed some 700 civilians. This is 50 civilians for every militant killed, a hit rate of 2 percent—hardly 'precision.'"[158] Kilcullen, who was an influential adviser to Gen. David Petraeus in Iraq, then went before the House Armed Services Committee and further legitimized the sourceless statistics: "I realize that [the drones] do damage to the Al Qaeda leadership. Since 2006, we've killed 14 senior Al Qaeda leaders using drone strikes; in the same time period, we've killed 700 Pakistani civilians in the same area. The drone strikes are highly unpopular. They are deeply aggravating to the population. And they've given rise to a feeling of anger that coalesces the population around the extremists and leads to spikes of extremism."[159]

As the wildly inflated, sourceless claims of the drones' civilian-to-militant kill ratio traveled from the Pakistani media to U.S. politicians, it became an article of faith in many circles that the CIA drones hunted not Taliban and al Qaeda, but innocent Pakistani civilians. Typical of this viewpoint was Maulvana Sami Ulhaq of the Pakistani Jamiat e Ulema e Islam (the Community of Islamic Scholars, an extremist Pakistani Islamist party), who announced at a conference in Lahore that the U.S. drone attacks kill "dozens of innocent people daily."[160] Muhammad Ahmed of the popular Buzz Pakistan website similarly wrote, "USA did more than 100 drone attacks in Pakistan in the past 3 years, if you read news about these drone attack you will see that in these drone attack only 1% terrorists was killed and other 99% people who died in these attack was innocent civilians of Pakistan. 75% of them were 10 to 15 year old teenagers."[161] The *Pakistan Observer* similarly reported, "The US drones or the predator planes which have been on the killing spree in Pakistan's northern belt since August 2008 and have so far killed over fourteen hundreds people with the big majority as the innocent civilians (as admitted by the international watch dogs)."[162] A similar finding was made by Zeeshan-ul-hassan Usmani, a Pakistani who runs a website called Pakistani Body Count.[163] His site, the only Pakistani site with sourcing, found that twelve hundred Pakistani civilians had been killed for forty terrorists. Usmani, however, came to this stunning conclusion by labeling all Taliban killed by drones as "civilians."[164] This even though the Pakistani government itself has recognized the members of Pakistani Taliban groups, who have killed thousands of Pakistani civilians, as terrorists. One Pakistani organization went even further than Usmani and claimed that 957 civilians had been killed in drone strikes in the year 2010 alone.[165]

Once again, these unfounded claims were not limited to Pakistan. A widely cited American article on civilian drone deaths by Daniel Byman of the Brookings Institution (without any sourcing or database to support his findings) claimed, "Critics correctly find many problems with this program, most of all the number of civilian casualties the strikes have incurred. Sourcing on civilian deaths is weak and the numbers are often exaggerated, but more than 600 civilians are likely to have died from the attacks. That number suggests that for every militant killed, 10 or so civilians also died."[166]

Among other Westerners who joined the antidrone frenzy was antiwar activist Cindy Sheehan (to be discussed in a later chapter), who raised the number of civilian deaths even higher: "The drone bombings . . . are killing at least a hundred or more innocent civilians for every so-called terrorist that they get. We think that this is morally reprehensible."[167] Likening drones to land mines and cluster bombs, one of Britain's most senior judges opined, "Unmanned drones that fall on a house full of civilians is a weapon the international community should decide should not be used."[168] Ted Rall, a writer for Commondreams.org, similarly wrote in an article titled "US Drone Planes Have a Nearly Perfect Record of Failure," "Civilized nations should band together to renounce and outlaw these sloppy and obscene aerial assassination attempts, which send the terrifying message that killing civilians is acceptable in the pursuit of justice."[169] A U.S.-based website that was used to rally protestors to a march against drone strikes at CIA Headquarters in Langley claimed, "The primary and proven case against drone attacks is that they pose a public danger that can only be deemed as indiscriminate bombing."[170]

It is easy to find antidrone comments from politicians, journalists, lawyers, bloggers, and activists who focus on the supposed mayhem being wreaked by drones on "thousands" of innocent Pakistani men, women, and children. The notion that robotic drones fly over Pakistani houses indiscriminately unleashing bombs on civilians has become familiar fodder for antiwar voices, as much a part of their discourse as CIA black sites, rendition, water boarding, and the Guantánamo Bay detention center.

There are many myths related to the drone campaign, but the greatest seems to be the notion that drones "invade" Pakistan and hover over the country's civilians, slaughtering them indiscriminately while avoiding striking their actual terrorist targets with their state-of-the-art technology. An exploration of some of the other stories related to the drones found in the subsequent chapter will help shed some light on the CIA's murky assassination campaign and expose some of the other myths related to it.

8

Spies, Lawyers, Terrorists, and Secret Bases

This is quite an awesome power, the power to label somebody as an enemy [then] wipe them out without judicial process of any kind.
—Jameel Jaffer of the American Civil Liberties Union (ACLU)

Many stories associated with the CIA's extraordinary drone assassination campaign in Pakistan shed light on this murky war. Following are only a few of them that bring to life some of the more interesting episodes.

THE SHAMSI AIR BASE EXPOSURE

One of the earliest myths about the drone campaign was that the CIA drones were flying from bases in Afghanistan to carry out their deadly missions. The Pakistani media and politicians frequently made references to "intrusions" or "violations" of Pakistani airspace by "Afghan-based" drones. In actuality, the majority of drones were flying from a remote airfield in Pakistan's Baluchistan Province, known as Shamsi, that had been given to the Americans after 9/11. This remote airstrip, which is 350 miles south of Waziristan, had originally been built by Arab sheikhs from the gulf states who used to fly to the region to hunt local birds. Its remoteness, proximity to the FATA, and distance from Taliban insurgents made it a perfect location for launching drone strikes with the Pakistanis' "don't ask, don't tell" agreement.

Although many in Pakistan suspected that the government was covertly supporting the drone strikes "from Afghanistan," even as it publicly condemned them, there seemed to be little official evidence of this until early

2009. The Pakistani government's double game was finally exposed on February 12, 2009, when Senator Dianne Feinstein, chair of the Senate Intelligence Committee, dropped a bombshell in a conference when she said, "As I understand it, these [drones] are flown out of a Pakistani base."[1]

American journalists treated the incident as if Feinstein had revealed a state secret, but media sources such as the BBC, the *New York Times*, and CNN had long before reported that the drones were based in Pakistan.[2] Even though Western media had previously reported on the Pakistani bases, Feinstein's public acknowledgment brought them to the attention of the Pakistani public for the first time.

In response, the Pakistani embassy in Washington announced, "There are no foreign bases in Pakistan."[3] Pakistani defense minister Ahmad Mukhtar similarly rebutted Feinstein's incautious remark and added, "We do have the facilities from where they can fly, but they are not being flown from Pakistani territory. They are being flown from Afghanistan."[4]

But the truth came out five days later, when London's *Times* published an article that featured satellite images obtained from Google Earth that clearly showed Predator drones on a runway in Shamsi.[5] Pakistan's *Dawn* subsequently reported the news and claimed, "The existence of drone bases inside Pakistan suggests a much deeper relationship with the United States on counter-terrorism than has been publicly acknowledged."[6] Shireen Mazari of the Pakistani *News* went even further and published a scathing opinion piece on the Shamsi revelation:

> What many of us had suspected seems to have now been revealed by no less a person than the Chairperson of the US Senate Intelligence Committee, Democrat Senator Dianne Feinstein—that US drones operating in Pakistan are in fact flown from an airbase in Pakistan. . . .
>
> Official sources have lost all credibility. After all, we have been officially briefed on more than one occasion that no drone flew without the knowledge of the Pakistani military. . . .
>
> The brazenness with which the government has chosen to lie not only to its people but to Parliament shows how little it cares for either. . . . In retrospect it is a sick mind that will continue to harp publicly on how the drone attacks are encouraging extremism and must be stopped

while covertly there has always been a Pakistani acquiescence to these drone violations of our sovereignty.[7]

Tempers were soothed somewhat when it was revealed that the United States had created a base in Jalalabad, Afghanistan, near the Pakistani border, to fly drones into the FATA. Many drones were shifted to Jalalabad, where they were said to take off almost hourly. The issue of the secret CIA drone base at Shamsi was, however, brought up again in 2009, when the *New York Times* reported that the CIA had hired contractors from the firm Xe Services (formerly known as Blackwater) to guard the base's perimeter and load bombs and missiles onto the drones.[8] Xe/Blackwater had gained great international notoriety after several of its armed contractors in Baghdad gunned down seventeen civilians in 2007. News of the massacre made its way to Pakistan and other Muslim countries in the region. When Pakistanis heard that contractors from the notorious company were operating in their country at Shamsi, there were howls of rage, and the CIA was forced to end its contract with Xe in 2009.

Shamsi came up again in June 2011, when the Pakistanis announced that they were closing the base to punish the United States for withholding promised funds.[9] The United States, however, announced that it had already ceased operations at the base and transferred them across the Afghan border to the base at Jalalabad. But this move did not seem permanent, and in December 2011 the Pakistanis again announced they were closing the base after U.S. Apaches and AC-130 gunships accidentally killed twenty-four Pakistani border troops during a firefight in a place called Salala. Regardless, the existence of secret CIA bases and Xe contractors on Pakistani soil created a sense among many Pakistanis that the United States was attempting to occupy their country.

THE ARREST OF RAYMOND DAVIS AND THE DATTA KHEL STRIKE OF MARCH 17, 2011

Rumors of CIA and Xe agents running around Pakistan became even more widespread after a CIA contractor named Raymond Davis was arrested for gunning down two Pakistanis trying to rob him in the eastern city of Lahore on January 27, 2011. Subsequent reports indicated that the two men killed were actually ISI spies tracking Davis.[10] When the Pakistanis arrested Davis, an ex–Special Forces soldier, they found a *pathrai* drone tracking chip on him. He was also found to have traveled to the FATA region twelve times without official permission.[11] If this were not enough, his mobile phone was found to have made calls to Waziristan.[12] The Pakistani media reported that Davis's main task was "to keep CIA network intact in the tribal agencies as well as Khyber Pakhtunkhwa."[13] One Pakistani official said, "This is not the work of a diplomat. He was doing espionage and surveillance activities."[14] For their part, the Americans claimed that Davis was a "diplomat" and therefore had immunity.

As a result of the embarrassing affair, which led to widespread anti-American protests and exposed the CIA's covert operations in Pakistan, the Pakistani government demanded that hundreds of U.S. operatives in the country leave. Tensions rose between Islamabad and Washington, and in an effort to placate the Pakistanis during the nearly two-month period that Davis was in custody, the CIA called a monthlong halt to the drone strikes. A Taliban commander said of Davis's arrest and the resulting lull, "The arrest of this guy is a very positive thing for us. Our forces used to be hit by attacks every other day. Now we can move more freely."[15]

Davis was released on March 16, 2011, following a *diyya* (blood money) payment of $2.3 million to the slain robbers' families, and the drone strikes began in force the very next day. The drone attack in Waziristan on March 17 had been strongly opposed by the new U.S. ambassador to Pakistan, Cameron Munter, who felt that a strike coming so close on the heels of Davis's release was insensitive and would infuriate the Pakistanis.[16] The CIA, however, overrode Munter's concerns, and the ensuing strike turned out to be a greater mistake than Ambassador Munter could have predicted.

The March 17 strike killed between twenty-six and forty-four people in Datta Khel, a Taliban-controlled village in North Waziristan.[17] Although news sources initially claimed that those killed in the strike were "militants," the *New York Times* subsequently reported that as many as fifteen of the people killed in the strike were actually tribal elders.[18] Others killed in the strike were described as merchants and tribal police. It became apparent that the elders had been engaged in a *jirga* designed to settle a dispute between two tribes over a chromite mine. Since the territory in question was controlled by the Taliban, a high-ranking commander loyal to Taliban leader Hafiz Gul Bahadur was officiating at the *jirga*. While the elders and Taliban militants were meeting, as many as four missiles slammed into the *jirga*, killing the high-ranking Taliban leader, ten of his followers, and numerous civilians. A BBC account of the Datta Khel strike reported, "Officials say two drones were involved. One missile was fired at a car carrying suspected militants. Three more missiles were then fired at the moving vehicle, hitting it and the nearby tribal meeting, or jirga."[19]

Regardless of the intended target, the local population was infuriated by the deaths of so many respected elders at the hands of the CIA. One survivor said, "It wasn't a militant gathering, but a meeting of tribal elders from Ismail Khan village to sort out some differences over a business deal. One of Bahadur's commanders, Sharabat Khan, was also present at the meeting as he is also a local elder, but they were discussing business." The surviving elders demanded blood money from the Americans for their slain family members, and one elder said that the attack "will create resentment among the locals and everyone might turn into suicide bombers." Another elder who survived said, "We are a people who wait 100 years to exact revenge. We never forgive our enemy."[20]

A Pakistani government source said, "The Pakistan Army condoles with the families whose dear and near ones have been martyred in this senseless attack." But an American official caustically rejected Pakistan's account of the civilian deaths and claimed those who were killed in the strike were Taliban militants: "These people weren't gathering for a bake sale. They were terrorists."[21] Another U.S. source in a similar sarcastic tone said, "There's every indication that this was a group of terrorists, not a charity car wash in the Pakistani hinterlands."[22]

Pakistan ignored the Americans' remarks and subsequently showed its displeasure by pulling out of tripartite talks between the U.S., Afghan, and Pakistan governments on the future of Afghanistan. The Pakistani Foreign Ministry issued the following statement: "It is evident that the fundamentals of our relations need to be revisited. Pakistan should not be taken for granted nor treated as a client state."[23]

THE TALIBAN KILL A CIA DRONE TEAM AND TRY TO SET OFF A BOMB IN TIMES SQUARE

The people of the FATA, and Pakistan in general, have many myths and legends about the *bangana* (thunder), *ghangais* (buzzers), or *machays* (wasps), as the drones are commonly known. When the rumor spread that the CIA had placed *pathrai* homing beacons in Lipton tea bags, for example, many Pashtuns in the region stopped drinking tea.

One of the more enduring myths is that the wave of suicide bombings that have swept Pakistan in recent years is directly related to the drone strikes. According to this myth, if there were no drone strikes, there would be no suicide-bombing slaughter of thousands of innocent Pakistanis. Such fears have been deliberately stoked by Taliban leaders, who have threatened to launch two suicide bombings per drone strike.[24]

I have been able to find only a few cases of terrorist attacks that appeared to be directed in response to the drone strikes. The first was the previously mentioned case of a suicide bombing on a Pakistani military base in November 2006, just days after the infamous drone strike killing eighty-one students and militants in the Chenagai suburb of Damadola.[25] Suicide bombings were still rather rare in Pakistan at that time, and the timing of the attack, so soon after the Chenagai strike, seemed to be in fulfillment of the madrassa head's promise of revenge.

Then there was the case of the failed Times Square bomber Faisal Shahzad. On May 1, 2010, Shahzad drove an SUV packed with explosives into the middle of Times Square in New York City and set off a timer on the bomb. Several observant bystanders, however, noticed smoke coming from the truck and called the police, who then defused the bomb. A fleeing Shahzad was subsequently arrested on a plane waiting for takeoff to Dubai.

It later emerged that Shahzad, who was a Pakistani American, had been trained in a terrorist training camp in Waziristan of the very sort the CIA drones had been attacking. Shahzad claimed to have been in the region at the time of CIA drone attacks and to have known people who were killed in drone strikes.[26] During his court trial Shahzad was questioned by Judge Miriam Cedarbaum. Following is an account of that questioning that shows how the drone strikes motivated him:

"You wanted to injure a lot of people," said Cedarbaum. Shahzad said the judge needed to understand his role. "I consider myself a Muslim soldier," he said. When Cedarbaum asked whether he considered the people in Times Square to be innocent, he said they had elected the U.S. government.

"Even children?" said Cedarbaum.

"When the drones hit, they don't see children," answered Shahzad. He then said, "I am part of the answer to the U.S. killing the Muslim people."[27]

Shahzad also proclaimed his desire to avenge "those innocent people being hit by drones from above."[28] He subsequently responded to the question of whether he wanted to plead guilty by saying, "I want to plead guilty, and I'm going to plead guilty 100 times over. Because until the hour the U.S. pulls its forces from Iraq and Afghanistan, and stops the drone strikes in Somalia and Yemen and in Pakistan, and stops the occupation of Muslim lands, and stops killing the Muslims, and stops reporting the Muslims to its government, we will be attacking U.S., and I plead guilty to that."[29]

The Pakistani Taliban subsequently claimed responsibility for Shahzad's failed terrorist attack, their second attempted bombing outside of the Afghanistan-Pakistan theater (the first being an attempt in Spain). In a video statement the Pakistani Taliban said it was revenge for the "recent rain of drone attacks in the tribal areas."[30] A Taliban commander said of the attempted terrorist attack, "We were expecting this. They were desperately looking for revenge against America inside America."[31] An American intelligence official concurred and said, "Those [drone] attacks have made it personal for the Pakistani Taliban—so it's no wonder they are beginning to think about how they can strike back at targets here."[32]

Although Hakimullah Mehsud failed in his effort to avenge the death of his predecessor, Baitullah, on this occasion, he succeeded in another attempt: the infamous Camp Chapman bombing, an account of which reads like a story in a spy novel. (It was actually featured in the Hollywood blockbuster *Zero Dark Thirty*.) The Camp Chapman bombing actually has its origins in Jordan, where police had arrested a Jordanian doctor named Hamam al Balawi, who had been involved with extremist groups. The Jordanian General Intelligence Directorate turned Balawi and made him a double agent. Convinced that he was now working for them against the terrorists, the Jordanians offered him to their American allies. Their hope was that Balawi would travel to Waziristan and offer his services to al Qaeda. After being accepted, he could then reveal Ayman al Zawahiri's location to the CIA, which could dispatch him with a drone.

Balawi began to work secretly for a CIA drone intelligence team based in Forward Operating Base Camp Chapman in the Afghan border province of Khost. There this "superstar asset" appeared to gain the trust of the CIA team handling him, so much so that he was once allowed on the base without being frisked. This lack of vigilance cost the CIA heavily on December 30, 2009. As Balawi arrived on the base, he was surrounded by CIA officers and contractors. When one of them belatedly went to pat him down, Balawi triggered a bomb hidden on his body and killed everyone around him, including the CIA station chief, a forty-five-year-old mother of three; two officers involved in the CIA Counterterrorism Center's drone-targeting program; two contractors for Xe/Blackwater; one Jordanian intelligence officer, who was a cousin of the Jordanian king; and one other person. Virtually the entire CIA drone team at Camp Chapman was wiped out by Balawi, who was actually a triple agent still working for the terrorists. Today a photo tribute to the slain agents hangs in the CIA Counterterrorism Center's Pakistan Afghanistan Department at Langley.[33]

After Balawi's suicide, Pakistani Taliban leader, Hakimullah Mehsud, released a prerecorded videotape that featured images of him sitting next to Balawi as the suicide bomber posthumously promised to avenge the earlier drone assassination of Baitullah Mehsud. Balawi declared, "We will never forget the blood of our emir, Baitullah Mehsud. We will always demand revenge for him inside America and outside."[34] Balawi also said he was offered "millions of dollars" to "spy on mujahideen . . . but instead I came to them [the Taliban] . . . and I told them everything, and we arranged this attack so the Americans can understand that the belief of Allah. . . . This jihadi attack will be the first of the latest operations against the Americans and their drone teams outside the Pakistan border, after they killed the Emir of Tehrik-i Taliban Pakistan Baitullah Mehsud."[35]

The Camp Chapman suicide attack was the second largest loss of CIA lives after the bombing of the U.S. embassy in Beirut in 1983, which destroyed an entire CIA team. Revenge was not, however, long in coming. After the Khost bombing, an intelligence official promised, "Last week's attack will be avenged. Some very bad people will eventually have a very bad day."[36] In the following weeks the CIA launched an unprecedented blitz of drone strikes on the territories of Mehsud and Haqqani, who was also thought to be involved in the Camp Chapman attack. Among those killed in the retaliatory strikes was Hussein al Yemeni, a top al Qaeda leader who was involved in the Camp Chapman plot. For all the CIA's efforts to take out both Haqqani and Mehsud, to date neither of them has been killed and the real culprits behind the Camp Chapman attack have gone unpunished.

THE EXPOSURE OF THE PAKISTANI CIA STATION CHIEF'S IDENTITY

In 2010 a Pakistani journalist named Kareem Khan from Mir Ali, North Waziristan, filed a lawsuit against the CIA station chief in Pakistan, CIA director Leon Panetta, and Secretary of Defense Robert Gates. Khan's suit asked for a half million dollars in compensation for the killing of his brother and his son in a drone strike on December 31, 2009. Khan claimed, "That drone attack killed my son, my brother and a local man. We are not terrorists, we are common citizens."[37]

But that was not all. In his lawsuit Khan named the head of the CIA in Pakistan, Jonathan Banks, and blew his cover. Khan's application to register the case stated, "Jonathan Banks is operating from the American Embassy in Islamabad, which is a clear violation of diplomatic norms and laws, as a foreign mission cannot be used for any criminal activity in a sovereign state."[38] Khan's lawyer further stated, "Mr. Kareem maintains that Jonathan Banks is not a U.S. diplomat therefore he does not enjoy diplomatic immunity, and his involvement in the execution of his son and brother simply makes him a murderer who is to be taken to task."[39] He also called for Banks to be tried for murder and executed.

In response, the Islamabad Police Department moved to order a murder case against Banks. Within days Banks's name was published in papers around the globe, and protestors in Pakistan were carrying placards with his name on them demanding his arrest. Banks was hurriedly smuggled out of Pakistan after he received numerous death threats.

Enraged American officials claimed that Khan, a simple Pashtun tribesman, could not have discovered the name of the CIA station chief in Pakistan if he had not been given details of his identity by the Pakistani ISI.[40] As for the actual lawsuit itself, Khan has received no compensation money from the government to date, but his suit did achieve another purpose, namely, it embarrassed the CIA. The basis for Khan's lawsuit—that he was an innocent tribesman whose loved ones were brutally killed in a drone strike—was, however, challenged when two intelligence officials told CNN that he had been housing a notorious Taliban commander named Haji Omar Khan at the time of the strike.[41] Haji Omar Khan was said to have been killed in the drone strike.

THE ARREST OF THE "CREECH 14" AND OTHER ANTIDRONE PROTESTS IN THE UNITED STATES AND UK

As the drone strike campaign picked up in the final years of the Bush administration and then skyrocketed under Obama, an antidrone movement appeared in the United States. One of the most active components of this antidrone/antiwar movement was a women's group known as Code Pink. Code Pink recruited well-known activist Cindy Sheehan, who famously camped near President George W. Bush's home in Crawford, Texas, to protest the Iraq War following the death of her son in that conflict. Sheehan was by 2008 nationally known and began a "No Drones" bus tour in which she and other activists dressed in pink protested outside Air Force bases that were home to drone squadrons. The women also protested at CIA Headquarters in Langley and marched from there to former Vice President Dick Cheney's house, also in Langley, calling for his arrest. Code Pink protested at the San Diego home of the General Atomics chief executive officer, James Blue, where they built a shrine to the children killed by the drones his company produced. A statement on the Code Pink website concerning the protests read,

> So many civilian casualties by US drone attacks are not just war crimes but crimes against humanity. We must restore America's image by suspending these attacks immediately. This indiscriminate method of drone killing will not improve our relationship with Pakistan or the

Muslim world. It will not bring us safety or peace throughout the world, in fact it begets more harm and destruction as extremists use the death of innocent civilians as a tool to recruit more people to join the Taliban to fight against us.[42]

On her own website Cindy Sheehan wrote, "The primary and proven case against drone attacks is that they pose a public danger that can only be deemed as indiscriminate bombing."[43]

Thirty members of Code Pink, dressed in pink shirts that said "Stop Killer Drones," partook in an antidrone march led by Pakistani politician Imran Khan, an ex-cricketeer who has been vocal in his calls for peace with the Taliban militants and an end to the drone campaign against them. The October 2012 march of several thousand Pakistanis and Code Pink members made its way from Islamabad to South Waziristan before it was halted by the Pakistani army at the border because of Taliban security threats. There protestors (presumably not the Americans) chanted "Down with America" and "A Friend of America Is a Traitor to the Nation."[44] While in Pakistan, the members of Code Pink also delivered a protest letter condemning the drone strikes to the U.S. embassy in Pakistan; the letter was signed by, among others, actor Danny Glover, Massachusetts Institute of Technology professor Noam Chomsky, and movie director Oliver Stone. One of the American drone protestors described Obama's role as "chief executioner" at a Pakistani seminar on the issue.[45]

Another protest was held in October 2011 in Washington, D.C., at the Smithsonian's National Air and Space Museum, which had an exhibit dedicated to drones. As many as two hundred protestors with signs that read "Drones Kill Kids" were met with pepper spray and arrests when they tried to close down the exhibit.[46] A smaller protest was held outside a Raytheon Missiles System plant in Tucson, Arizona. There protesters held placards that read "We Have Guided Missiles and Misguided Men" and "Drone Attacks Inspire Hatred of the U.S."[47]

Perhaps the most famous antidrone protestors, however, were the "Creech 14," members of the antiwar group Nevada Desert Experience whose motto was "Ground the Drones Lest We Reap the Whirlwind." The Nevada Desert Experience was opposed to the "insidious creep of robotics into warfare." Fourteen of its members were arrested for trespassing on Creech Air Force Base near Las Vegas, Nevada, in 2009. Creech was the most famous base associated with the U.S. Air Force's separate drone campaign, which was largely carried out in the Afghan and Iraqi theaters of action. The group later summed up its actions at Creech as follows: "Nonviolent resisters want the U.S. government, the Pentagon, the drone controllers and the general populace to think about the horrific death and destruction the unmanned aerial attacks are raining down on people thousands of miles away and to contemplate that these attacks do not prevent or eliminate terrorism, but instead incite more hatred, revenge and retaliation, and make more recruits for the Taliban. . . . Warfare is not a video game."[48] The judge in the case against the Creech 14 found them guilty of trespassing but released them for time already served after telling them to "go in peace" and to use diplomacy instead of trespassing in the future to make their point.

In April 2011 thirty-eight protestors were similarly arrested at Hancock Field near Syracuse, New York, another base from which pilots of the 174th Fighter Wing remotely flew drones in Afghanistan and Iraq. Several of the protestors covered themselves in white sheets that had been painted red to resemble blood.[49] In this area protests were led by Christians who were opposed to the remote killing. One antidrone blogger in the area wrote,

To sit at a console 7,000 miles away with life and death control over people whose land you've never walked on is too much power for any human being. It makes killing virtual and is a virtual license to kill. It can only corrupt.

I call on every pastor and minister in the Syracuse area to begin each service with an apology to the children of South Waziristan for the terror we have inflicted in their skies.[50]

Another antidrone voice submitted an article to the *Buffalo News* that read, "Drone attacks are extrajudicial executions, with pilots acting as detective, prosecutor, judge, jury and executioner. There is no due process or rule of law, and the government claims the right to apply this policy to American citizens abroad."[51]

Among the antidrone voices has been U.S. congressman Dennis Kucinich, who wrote in the *Huffington Post*, "Think of the use of drone air strikes as summary executions, extra-judicial killings justified by faceless bureaucrats using who-knows-what 'intelligence,' with no oversight whatsoever and you get the idea that we have slipped into a spooky new world where joystick gods manipulating robots deal death from the skies and then go home and hug their children."[52]

Peter Singer, the author of a book on military robotics titled *Wired for War*, similarly reflected on drones in a *New York Times* article titled "Do Drones Undermine Democracy?"

> Now we possess a technology that removes the last political barriers to war. The strongest appeal of unmanned systems is that we don't have to send someone's son or daughter into harm's way. But when politicians can avoid the political consequences of the condolence letter—and the impact that military casualties have on voters and on the news media—they no longer treat the previously weighty matters of war and peace the same way.
>
> For the first 200 years of American democracy, engaging in combat and bearing risk—both personal and political—went hand in hand. In the age of drones, that is no longer the case.[53]

An *Economist* article read, "Looking farther ahead, there are fears that UAS [unmanned aerial systems] and other robotised killing machines will so lower the political threshold for fighting that an essential element of restraint will be removed. Robert E. Lee said 'it is well that war is so terrible, otherwise we would grow too fond of it.' Drones might make leaders fonder of war."[54] Amnesty International similarly warned, "Anything that dehumanizes the process makes it easier to pull the trigger."[55]

The antidrone sentiment was not limited to the United States and Pakistan. In October 2011 a "week of action" was called for in Britain by a group named Ground the Drones. This group led small protests at various sites in the UK linked to drones. (The U.S. military has leased several Reaper drones to the British military for use in combat in Afghanistan.) These sites included the London office of the drone manufacturer General Atomics; the

Royal Air Force base at Northwood, where the drone pilots were based; Boscombe Down, the testing ground for a domestic surveillance drone used by the police and known as the Watchkeeper; and other venues.

British police also arrested twenty members of the United Ummah (*ummah* translates as "Islamic community") who were protesting U.S. drone strikes outside the U.S. embassy in London. The group claimed to be protesting against "the recent spate of anti-Muslim drone strikes that have been launched by the U.S. government against innocent Muslims."[56]

As the use of drones increases in both the United States, where they have been used for surveillance on the Mexican border and in an arrest in North Dakota, and the UK, news of the expanded use of drones overseas in Pakistan, Afghanistan, Iraq, Yemen, Libya, and Somalia has been mainstreamed by the UK and U.S. media. As a result, public awareness of the drones has increased, and this has led to mounting opposition to the world's first "robotic" assassination campaign. (The drones are not actually autonomous robots, of course; humans still control them.) Antidrone activists, for example, appeared among the Occupy Wall Street protestors in the fall of 2011. One of them said, "The army considers human losses as collateral damage. Drone attacks are not just killing particular people but they are bombing entire villages and killing innocent civilians. It's absolutely horrible that we even have that technology."[57]

The issue that many protestors find galling is, ironically enough, the very aspect of the drones that leads to their unique precision, i.e., the fact that they are flown remotely by pilots who use cameras to closely monitor their targets. Opponents fear that the drones represent a slippery slope down the path to the "roboticization" of warfare and the rise of a sanitized video-game mentality of assassination by "kill TV." The very remoteness of the pilots from danger, they fear, will lead to an increased reliance on drones to kill real or suspected terrorists throughout the world. Soon other countries will be turning their suspected terrorists into "bugsplats" with drones. These activists fear a future when modern states use the drones to kill their enemies across the globe without trials. One antidrone activist summed up this sentiment when he wrote, "These silent surreptitious killings seem antiseptic, somewhat like playing a video game. We do not hear about the horror and grief these executions create. It will be all too easy for the citizens of this country to sanction these kinds of extra-legal activities when they know nothing about their devastating consequences. The absence of U.S. lives lost may make it easier for us to enter wars in the future."[58]

Thus the world's first remote-control assassination/bombing campaign has led to the rise of the world's first antidrone campaign, which shows no sign of weakening. As the U.S. military and intelligence agencies increasingly rely on drones in the war against al Qaeda in Pakistan, Somalia, and Yemen, the opposition is bound to increase exponentially.

THE BATTLE OF THE DRONE LAWYERS

One of the most fascinating issues of the CIA's drone campaign, which began as a targeted assassination campaign and then expanded into an all-out war, is its legality. Does the CIA have the right under both international and domestic law to kill thousands of people in a country that the United States is not officially at war with, or has it gone down a dangerous path toward legitimizing a cross-border campaign of mass extrajudicial killings?

These important questions were raised in 2009 and 2010 by Philip Alston, a professor of law at New York University who also served as the UN's special rapporteur on extrajudicial killing and arbitrary executions. In an extensive official report to the UN, Alston wrote that America's "strongly asserted, but ill-defined license to kill without accountability is not an entitlement which the United States or other states can have without doing grave damage to the rules designed to protect the right to life and prevent extrajudicial executions." The report further said, "This expansive and open-ended interpretation of the right to self-defense goes a long way towards destroying the prohibition on the use of armed force contained in the U.N. Charter. If invoked by other states, in pursuit of those they deem to be terrorists and to have attacked them, it would cause chaos." The report ended by saying, "Because operators are based thousands of miles away from the battlefield, and undertake operations entirely through computer screens and remote audio-feed, there is a risk of developing a 'PlayStation' mentality to killing."[59]

In an earlier report to the UN, Alston wrote that it was difficult to make any real assessments on the legality of the CIA's unprecedented drone campaign because it was veiled in secrecy. He called on the U.S. government to provide more information on how targets were selected and civilian casualties avoided. He also called on the United States to explain the legal basis for its unprecedented drone campaign and to outline how it complied with humanitarian law. If the United States did not do so, Alston warned, it would "increasingly be perceived as carrying out indiscriminate killings in violation of international law." He also said, "The onus is really on the government of the United States to reveal more about the ways in which it makes sure that arbitrary executions, extrajudicial executions are not, in fact, being carried out through the use of these weapons."[60] Alston added, "Otherwise you have the really problematic bottom line, which is that the Central Intelligence Agency is running a program that is killing significant numbers of people and there is absolutely no accountability in terms of the relevant international laws."[61]

The Obama administration was clearly taken aback by Alston's broadside and felt that some response was necessary to demonstrate that the drone campaign was a legal, responsible, and appropriate response to the terrorist threat that the United States was facing. Harold Koh, legal adviser to the U.S. State Department, was chosen to make that response. In March 2010 Koh gave a speech in Washington, D.C., to the Annual Meeting of the American Society of International Law that for the first time laid out the Obama administration's rationale for relying on drones:

> With respect to the subject of targeting, which has been much commented upon in the media and international legal circles, there are obviously limits to what I can say publicly. What I can say is that it is the considered view of this Administration—and it has certainly been my experience during my time as Legal Adviser—that U.S. targeting practices, including lethal operations conducted with the use of unmanned aerial vehicles, comply with all applicable law, including the laws of war.
>
> The United States agrees that it must conform its actions to all applicable law. As I have explained, as a matter of international law, the United States is in an armed conflict with al-Qaeda, as well as the Taliban and associated forces, in response to the horrific 9/11 attacks, and may use force consistent with its inherent right to self-defense under international law. As a matter of domestic law, Congress authorized the use of all necessary and appropriate force through the 2001

Authorization for the Use of Military Force (AUMF). These domestic and international legal authorities continue to this day.

As recent events have shown, al-Qaeda has not abandoned its intent to attack the United States, and indeed continues to attack us. Thus, in this on-going armed conflict, the United States has the authority under international law, and the responsibility to its citizens, to use force, including lethal force, to defend itself, including by targeting persons such as high-level al-Qaeda leaders who are planning attacks. As you know, this is a conflict with an organized terrorist enemy that does not have conventional forces, but that plans and executes its attacks against us and our allies while hiding among civilian populations. That behavior simultaneously makes the application of international law more difficult and more critical for the protection of innocent civilians.[62]

Alston subsequently issued a rebuttal to Koh's statement: "I don't think we can ask for full transparency. I don't think the United States is ever going to provide access to all of the information relating to these killings. But until it starts to provide at least some access, we will not be able to conclude that the United States is in fact complying with the law, as Harold Koh insisted."[63]

Koh did not respond to Alston's continued requests for the release of further details. But the *Washington Post* published an article in support of Koh's statement, which said, "[Koh] rightly rejected the absurd notion that enemy targets must be provided 'adequate process' before the strike occurs. . . . Mr. Koh's reaffirmation of the right to self-defense—even outside the confines of an existing armed conflict—is particularly important. . . . The right of self-defense is inherent and may be exercised against current and future enemies that pose an imminent threat, including those operating outside of traditional combat zones."[64]

Regardless of the criticisms from Alston, the Obama administration has continued to refer to the 2001 AUMF signed by President George Bush, which gave the president the authority "to use all necessary and appropriate force against those nations, organizations, or persons he determines planned, authorized, committed, or aided the terrorist attacks that occurred on September 11, 2001, or harbored such organizations or persons."

There have been several other interesting legal cases involving drones as well. One of them concerns a forty-year-old Yemeni American named Anwar al Awlaki. Awlaki grew up in Las Cruces, New Mexico, but became devoted to jihad against the United States after 9/11. He moved to Yemen, linked up with al Qaeda in the Arabian Peninsula (AQAP), and became an outspoken supporter of jihad against his former American homeland. His e-mails and online sermons in Arabic, in which he called Americans "devils," proved to be inspiration for Nidal Hasan, another American Muslim who went on a shooting spree at the U.S. army base at Fort Hood, Texas, that left thirteen dead. After Hasan's attack Awlaki said, "Nidal Hasan is a hero."[65]

Awlaki's fiery preaching was also said to have inspired Faisal Shahzad, the failed Times Square bomber.[66] In the online jihad magazine *Inspire*, Awlaki called for the killing of American civilians: "The ideal location [for a bombing] is a place where there are a maximum number of pedestrians and the least number of vehicles. In fact if you can get through to 'pedestrian only' locations that exist in some downtown [city center] areas, that would be fabulous."

Awlaki's fluent American English and knowledge of the culture and idioms of the West made him an ideal recruiter for disaffected Muslims, like Nidal Hasan, who were living in America. He was called the "bin Laden of the Internet." But Awlaki proved to be more than just an inspiration. President Obama called him the "leader of external operations for Al Qaeda in the Arabian Peninsula."[67] On another occasion he said that Awlaki "took the lead in planning and directing efforts to murder innocent Americans."[68] Specifically, he was actively involved in the December 25, 2009, "underwear bomber" plot to blow up an airliner flying from Amsterdam to Detroit.[69] He was also accused of involvement in an attempt to blow up a United Parcel Service (UPS) plane flying from Yemen using bombs triggered by a cell phone. If this were not enough, he was found to have been plotting with a British Muslim of Bangladeshi decent to blow up a British airliner.[70]

In response to his increasingly dangerous operational role, the CIA announced that it had added Awlaki to a terrorist hit list; he was the first American to earn this distinction. At the time a secret memo that stated that it would be lawful to kill Awlaki only if he could not be taken alive was written. This memorandum was written after meetings in the White House Situation Room involving lawyers from the State Department, Pentagon,

National Security Council, and various intelligence agencies. The secret document provided the justification for killing Awlaki, despite an executive order banning assassinations and a federal law against murder. The Justice Department decided that the killing of Awlaki, who was a member of a "cobelligerent" group (AQAP), was covered by the AUMF. The *New York Times* called the memo, which legitimized the killing U.S. citizen without a trial, "one of the most significant decisions made by President Obama."[71]

In February 2013 a Justice Department memo to the Senate Intelligence and Judiciary Committees that outlined the legal argument for the strikes against U.S. citizens, such as Awlaki, was leaked to NBC. The sixteen-page memo provided guidelines for killing Americans abroad if they were senior members of a terrorist organization and involved in operations against the United States. The memo stated that it is "lawful for the US to target al-Qaeda-linked US citizens if they pose an 'imminent' threat of violent attack against other US citizens, and that delaying action against such people would create an unacceptably high risk."[72] The memo required that the capture of American citizens involved in terrorism had to be deemed unfeasible before lethal operations could be carried out against them. Awlaki, who was involved in ongoing terrorist operations against U.S. citizens and guarded by pro–al Qaeda tribesmen in a remote corner of Yemen, certainly fit the criteria covered in the memo.

The United States subsequently pressured the Yemenis to arrest this major al Qaeda recruiter and planner. On one occasion Yemeni troops surrounded a village where Awlaki was hiding out with tanks, but he managed to escape. The JSOC subsequently made several well-publicized attempts to kill him with drones in Yemen.[73]

These developments caused considerable controversy in the United States. The idea that the U.S. government could execute a U.S. citizen with no judicial process based on secret intelligence made many uneasy.[74] One counterterrorism official, however, responded to criticism by saying, "American citizenship doesn't give you carte blanche to wage war against your own country. If you cast your lot with its enemies, you may well share their fate."[75] The *Wall Street Journal* similarly opined, "Perhaps al-Awlaki's U.S. citizenship gave U.S. officials pause, but after he joined the jihad he became an enemy and his passport irrelevant."[76]

In response to these extraordinary developments, Awlaki's father, Naseer Awlaki, sued CIA director Leon Panetta and Defense Secretary Robert Gates with the support of the ACLU. The senior Awlaki described his son somewhat implausibly as an "all-American boy." The ACLU lawsuit claimed that the Obama administration had given itself "sweeping authority to impose death sentences on American citizens."[77] In December 2010, however, a federal judge threw out the lawsuit claiming that Gates and Panetta were immune from lawsuits and that Awlaki's father did not have the standing to file the suit.[78] In essence the judge ruled that the court did not have the right to overrule President Obama (the executive branch) in wartime on matters involving war.

In March 2010 the ACLU filed a Freedom of Information of Act lawsuit that demanded that the CIA release information on the drone strikes. According to the ACLU website,

> The public has a right to know whether the targeted killings being carried out in its name are consistent with international law and with the country's interests and values. The Obama administration should disclose basic information about the program, including its legal basis and limits, and the civilian casualty toll thus far. . . .
>
> Recent reports, including public statements from the director of national intelligence, indicate that U.S. citizens have been placed on the list of targets who can be hunted and killed with drones.[79]

This suit was also thrown out in September 2011 by a federal judge who stated that the ACLU did not have the right to access information on the CIA's secret programs.

Then, on September 30, 2011, news came from Yemen that made headlines around the world. Awlaki and five companions had been tracked down and killed by a CIA Predator missile while they traveled across the north Yemeni desert in a truck. An eyewitness gave the following account of their assassination: "The witness said Awlaki was travelling in a pick-up with six other people on their way to neighboring Marib province. They stopped for breakfast in the desert and were sitting on the ground to eat when they spotted drones, so they rushed to their truck. A Hellfire missile fired from a drone struck the truck, leaving it a charred husk and killing all of those inside."[80]

Because his vehicle had been carbonized in the attack, Yemeni sources were "100% sure" Awlaki was dead. Killed alongside him was another American who had thrown in his lot with al Qaeda, the Pakistani-born Samir Khan. Khan was naturalized in his youth as a U.S. citizen and grew up in New York. He was the editor of *Inspire* and had once proclaimed, "I am proud to be a traitor to America."[81] Among other articles he published in the magazine was "How to Make a Bomb in the Kitchen of Your Mom."

Awlaki's assassination revived the controversy surrounding the drone strikes, and Jameel Jaffer, the ACLU's deputy legal director, said the killing violated U.S. and international law. He said, "As we've seen today, this is a program under which American citizens far from any battlefield can be executed by their own government without judicial process, and on the basis of standards and evidence that are kept secret not just from the public, but from the courts."[82] Samir Khan's family released a statement that asked, "Was this style of execution the only solution? Why couldn't there have been a capture and trial?"[83]

Libertarians and Muslim advocates further claimed that Awlaki and Khan had been denied their Fifth Amendment rights to protection from governmental legal abuse, including their right to due process. Republican presidential contender Congressman Ron Paul, a libertarian, said, "He was never tried or charged with any crime. . . . If the American people accept this blindly and casually that we now have an accepted practice of the president assassinating people who he thinks are bad guys, I think it's sad."[84]

His son, Senator Rand Paul, also a libertarian, would go even further by launching a thirteen-hour filibuster in March 2013 in an attempt to delay the hearing on Obama's nominee for CIA director, John Brennan. Paul said, "I will speak as long as it takes until the alarm is sounded from coast to coast that our Constitution is important. . . . That your rights to trial by jury are precious, that no American should be killed by a drone on American soil without first being charged with a crime, without first being found to be guilty."[85] Paul summed up his aims saying, "I'm here to filibuster John Brennan's nomination to be director of CIA. . . . I will speak for as long as it takes. I will speak today until the president says, 'no' he will not kill you at a café."[86] In a statement critics called fear mongering, but that was applauded by the Code Pink antidrone movement, Paul said, "That Americans could be killed in a café in San Francisco or in a restaurant in Houston or at their home in Bowling Green, Kentucky, is an abomination."[87]

Paul's comments were sharply criticized by Republicans senators Lindsey Graham and John McCain as alarmist and hypocritical considering his silence during Bush's drone campaign. Referencing the mounting criticism of Obama's drone campaign by Fox News and many Republicans, Graham said, "To my Republican colleagues, I don't remember any of you coming down here suggesting that President Bush was going to kill anybody with a drone, do you? They had a drone program back then, all of a sudden this drone program has gotten every Republican so spun up. What are we up to here?" He added, "People are astonished that President Obama is doing many of the things that President Bush did. I'm not astonished. I congratulate him for having the good judgment to understand we're at war. And to my party, I'm a bit disappointed that you no longer apparently think we're at war."[88] For his part McCain said, "We've done, I think, a disservice to a lot of Americans by making them think that somehow they're in danger from their government. They're not. But we are in danger from a dedicated, longstanding, easily replaceable-leadership enemy that is hellbent on our destruction."[89]

As the Paul filibuster demonstrated, no issue related to the drone campaign caused more consternation among Americans than the notion that drones could be used to kill Americans. Writing for the *Huffington Post*, columnist Anthony Gregory expressed the concerns of Paul and many other Americans in both parties over the precedent of an American president ordering the killing of an American without a trial:

> It was actually something special when President Obama ordered the drone killing of U.S. citizen Anwar al-Awlaki. . . .
>
> Obama's summary execution of these Americans, conducted as a military operation through the CIA, would indeed seem to break with precedent and qualify as one of those watershed moments in America's long retreat from the rule of law. While liberals criticized the Bush administration for warrant-less wiretapping and detentions without trial, one would think that the outright killing of an American citizen without due process would qualify as a greater offense.[90]

Journalist David Cole, writing for the *New York Review of Books*, expressed his concerns as well: "As American citizens we have a right to know when our own government believes it may execute us (and others) without a trial."[91]

For his part, President Obama said that Awlaki had been "planning and directing efforts to murder innocent Americans." His death demonstrated that terrorists "will find no safe haven anywhere in the world."[92] Most Americans seemed to agree with Obama. According to a subsequent poll, a full 69 percent of Americans who were asked if the strike on Awlaki was justified agreed that it was.[93]

To further gain legitimacy for the strikes, on March 5, 2012, Attorney General Eric Holder gave a talk on the issue at Northeastern University, saying that "the unfortunate reality is that our nation will likely continue to face terrorist threats that—at times—originate with our own citizens. We must take steps to stop them—in full accordance with the Constitution. In this hour of danger, we simply cannot afford to wait until deadly plans are carried out—and we will not." He further said, "Given the nature of how terrorists act and where they tend to hide, it may not always be feasible to capture a United States citizen terrorist who presents an imminent threat of violent attack. In that case, our government has the clear authority to defend the United States with lethal force."[94]

Holder continued, "Any decision to use lethal force against a United States citizen—even one intent on murdering Americans who has become an operational leader of Al Qaeda in a foreign land—is among the gravest that government leaders can face." He then laid out a three-part formula for determining whether the targeted killing against a U.S. citizen is legally justified. He said the government must determine that the citizen in question poses an imminent threat of violent attack against the United States, capture is not possible, and the killing would be consistent with laws of war.[95] The drone strikes against those posing an imminent threat would be limited to American terrorists abroad. In March 2013 Holder stated, "The government has no intention to carry out any drone strikes in the United States. It's hard for me to imagine a situation in which that would occur."[96]

As if to hammer home the point that the Obama administration would kill those Americans deemed to be terrorists operating with the enemy abroad, in the month following Awlaki's death a CIA drone also killed

Awlaki's sixteen-year-old son, Abdul Rahman, and a seventeen-year-old Yemeni cousin in a blitz of drone strikes on al Qaeda vehicles and safe houses in Yemen.[97] Abdul Rahman Awlaki's grandfather furiously denied the charges that his grandson was a terrorist and said, "To kill a teenager is just unbelievable, really, and they claim that he is an al-Qaeda militant. It's nonsense. They want to justify his killing, that's all."[98] For their part, U.S. officials claimed they had "no idea" the son was traveling with a supposed Egyptian al Qaeda agent, and a senior Obama administration official added, "This was a military-aged male travelling with a high-value target."[99]

Ironically enough, postings on several Yemeni al Qaeda websites after the son's death claimed that Abdul Rahman Awlaki, who had been dubbed "Usayyid" (the Lion's Cub) by al Qaeda after his father's death, had sworn that he wanted to be martyred just as his father had been before him. One al Qaeda site said, "His sadness reached its peak after the American planes assassinated his father. . . . But when he said to the Emir [al Qaeda leader] of the city of Azzam, 'I hope to attain martyrdom as my father attained it,' it did not come to his mind that this will happen, and just one day after he said it."[100]

OBAMA'S PERSONAL ROLE IN APPROVING A SECRET DRONE KILL LIST

On May 29, 2012, the *New York Times* published an extraordinary article by Jo Becker and Scott Shane titled "Secret 'Kill List' Proves a Test of Obama's Principles and Will," which reported that the president was intimately involved in the moral and targeting decisions made in the drone strikes in Yemen and Somalia.[101] The article depicted a president who realized after several close terrorist calls (including the failed bombing of a flight to Detroit in 2009 by the underwear bomber) that if even one terrorist attack succeeded, his presidency would be marked by it. According to Becker and Shane,

> Nothing else in Mr. Obama's first term has baffled liberal supporters and confounded conservative critics alike as his aggressive counterterrorism record. . . .
>
> While he was adamant about narrowing the fight and improving relations with the Muslim world, he has followed the metastasizing

enemy into new and dangerous lands. When he applies his lawyering skills to counterterrorism, it is usually to enable, not constrain, his ferocious campaign against Al Qaeda—even when it comes to killing an American cleric in Yemen [Awlaki], a decision that Mr. Obama told colleagues was "an easy one." ...

This secret "nominations" process [of names to add to the "kill list"] is an invention of the Obama administration, a grim debating society that vets the PowerPoint slides bearing the names, aliases and life stories of suspected members of Al Qaeda's branch in Yemen or its allies in Somalia's Shabab militia. The video conferences are run by the Pentagon, which oversees strikes in those countries, and participants do not hesitate to call out a challenge, pressing for the evidence behind accusations of ties to Al Qaeda. ... A parallel, more cloistered selection process at the C.I.A. focuses largely on Pakistan, where that agency conducts strikes.

The nominations go to the White House, where by his own insistence and guided by Mr. Brennan [national security adviser], Mr. Obama must approve any name. He signs off on every strike in Yemen and Somalia and also on the more complex and risky strikes in Pakistan—about a third of the total. Aides say Mr. Obama has several reasons for becoming so immersed in lethal counterterrorism operations. A student of writings on war by Augustine and Thomas Aquinas, he believes that he should take moral responsibility for such actions. And he knows that bad strikes can tarnish America's image and derail diplomacy.

DRONES FOIL A MUMBAI-STYLE TERRORIST ATTACK IN EUROPE

For those who support Obama's withering drone campaign, there is no better proof of drones' success in foiling terror strikes than their role in killing several German and British Muslims in North Waziristan who were planning a mass-casualty terrorist strike in Europe based on the model of one carried out in Mumbai, India, in November 2008. During the infamous Mumbai attack, members of the Pakistani jihad group Lashkar e Toiba roamed the streets of Mumbai raiding restaurants, hotels, a cinema, a Jewish center, a train station, and a college. In all 168 people were gunned down in cold blood or blown up by hand grenades as the terrorists spread carnage through the heart of India's largest city.

The Mumbai attack was praised by jihadists and terrorists throughout Pakistan and seemed to serve as inspiration for similar slaughters in Europe, according to Ahmed Siddiqui. Siddiqui was a thirty-six-year-old German of Afghan decent who was arrested in June 2010 as he attempted to fly out of Kabul International Airport to Germany. He was subsequently interrogated by the United States and its Coalition allies at Bagram Air Base. Coalition forces had previously noticed that a group of eleven radicals tied to the Al Quds Mosque in Hamburg, Germany (the same mosque attended by 9/11 suicide team commander Muhammad Atta), had mysteriously left the country in 2009. They then reappeared in training camps in North Waziristan. There the German-Muslim jihadists were filmed training with weapons and making threats against the West.

Initially, the German authorities thought the group of eleven extremists had traveled to the Afghanistan-Pakistan border to fight Coalition troops, but Siddiqui dropped a bombshell after his arrest. During interrogation he confessed to being a member of a terrorist team that was planning to attack civilian targets in Germany, France, and Britain. According to one U.S. official, "They were going to attack multiple centers in Europe over a few days. They were going to shoot the hell out of people, terrorize them."[102] Their aim was to punish these NATO countries for their role in the Afghan conflict, much as terrorists had punished the Spanish government for its role in Iraq with the 2004 Madrid bombings.

Bin Laden, Siddiqui claimed, had personally approved the plot.[103] British Muslims of Pakistani descent who formed a group known as the Islamic Army of Great Britain would attack the British targets, and German Muslims who were members of the Islamic Movement of Uzbekistan would carry out the attacks in Germany and France. A second German jihadist who was arrested and "other sources" verified Siddiqui's claims.[104]

Word of the plot was leaked in September 2010, and the British, German, and French governments went on high alert. Travel advisories were issued to Americans warning them not to travel to those countries. The Eiffel Tower was closed twice because of bomb scares, and even the British royal family was put under special protection.

By this time the CIA had learned that the British terrorists in North Waziristan were being led by a Pakistani-British Muslim named Abdul Jabbar. Jabbar had been selected to lead the Islamic Army of Great Britain in carrying out massacres against "soft targets" in the UK.[105] He and his brother had supposedly been chosen to lead the group in a Taliban meeting.[106] But before he could carry out his plans, Jabbar and three others were killed in a drone strike on September 8, 2010.[107] In fact, that September saw the highest number of drone strikes thus far as the CIA unleashed a barrage of twenty-two strikes trying to disrupt the plot by killing the British and German terrorists. On October 4, 2010, the drones finally caught up with the German terrorists, who were of Pakistani, Turkish, and Arab origin, and killed eight of them in the Mir Ali region of North Waziristan.[108]

Three months later two more Brits, this time English converts to Islam named Abu Bakr (aka Gerry Smith) and Mansoor Ahmed (known only as Stephen), were killed in a drone strike in the region.[109] It later emerged that CIA spies had been tracking the various German and British terrorists for some time, waiting for the chance to kill them and foil their plot. At the time U.S. officials said that they were acting on "precise intelligence."[110] The robust and accurate nature of the CIA's drone response to these terrorist threats speaks volumes to the level of their infiltration of the FATA and the agency's ability to quickly target and kill suspected terrorists planning future attacks.

DRONES AND THE KILLING OF OSAMA BIN LADEN

Although drones did not kill bin Laden in May 2011, they did play a key role in his death. Bin Laden had last been seen fleeing from the caves of Tora Bora in Afghanistan into the neighboring FATA region during Operation Enduring Freedom in December 2001. From there his trail ran cold. Many assumed that the terrorist mastermind was hiding in the FATA region because the number two in al Qaeda, Ayman al Zawahiri (who was tracked in the region and almost killed by a drone in Damadola), was filmed walking with him sometime after 2001 in a mountainous setting; both were wearing Pashtun clothes and interacting with people who appeared to be Pashtuns.

But by 2008 the drone war had commenced in the FATA, and many Taliban and al Qaeda leaders had fled this sanctuary for safer grounds in Quetta and elsewhere. Bin Laden himself was too high value a target to move around a region that was filled with CIA spies. He was forced to flee the drones by moving to the neighboring North-West Frontier Province. There he was isolated from the war in Afghanistan in a massive, $1 million, concrete compound built in the Pakistani military-resort town of Abbottabad.

It proved all but impossible to find bin Laden in his hideout because he was so far from the FATA. To compound matters, the terrorist leader had learned to be cautious following the 2004 death of Taliban leader Nek Muhammad, who, as noted previously, was tracked by the CIA while speaking on a cell phone. Instead of phones, bin Laden relied on couriers to communicate. For years bin Laden, occasionally issuing a video statement mocking the Americans, remained hidden in his self-imposed prison on the top floor of his compound.

But the CIA and NSA were persistent and carefully monitored the cell phone communication among all known al Qaeda agents in Pakistan. In August 2010 an al Qaeda agent tripped up and called a mysterious courier who had not previously been tracked named Ahmed al Kuwaiti. After the phone call, the NSA was tasked with following Kuwaiti's movements. Not realizing he was being traced, Kuwaiti drove to a large compound in Abbottabad that stood out because of its high walls. Curiosity aroused, the CIA began to monitor the mysterious mansion via high-flying spy satellites that used synthetic aperture radar and electro-optical cameras. The grainy pictures these satellites provided were not of sufficient resolution to tell who was living in the house. Although the satellites were good at filming static targets such as buildings, they could not film moving people; only drones could do that. Unfortunately, the Pakistanis did not allow the Americans' slow-moving propeller-driven Predators and Reapers to operate outside the FATA.

Around this time the CIA brought in a top-secret stealth drone known as the RQ-170 Sentinel that was jet propelled, shaped like a mini B2 bomber, and built to avoid radar. It was not armed but had the ability to penetrate undetected deep into Pakistani airspace to Abbottabad, a sensitive military town located near the Pakistani capital.[111] Soon after the mysterious compound was discovered, the drone was deployed to Abbottabad. According to the *Washington Post*, "Using unmanned planes designed to evade radar detection and operate at high altitudes, the agency conducted clandestine flights over the compound for months before the May 2 assault in an effort to capture high-resolution video that satellites could not provide. The aircraft allowed the CIA to glide undetected beyond the boundaries that Pakistan has long imposed on other U.S. drones, including the Predators and Reapers that routinely carry out strikes against militants near the border with Afghanistan."[112]

Having reached a 60–80 percent certainty that bin Laden was in the Abbottabad compound, the Obama administration, the CIA, and JSOC needed to formulate a plan to get him.[113] This was no easy task because the compound was a well-built concrete structure located both deep in the Islamabad Defense Intercept Zone and adjacent to a Pakistan military base. The slow-moving Reaper drones were only allowed to operate in "kill boxes" in the FATA and were not allowed to fly into Pakistan proper. Even if they had had a larger flight radius, their Paveway guided bombs were not strong enough to destroy bin Laden's compound. In addition, if the slow-moving, propeller-driven drones tried to penetrate the Islamabad security zone, they would be noticed by Pakistani radar and shot down. For these reasons, Obama and Gen. Bill McRaven, the head of JSOC, decided to use a dangerous helicopter-borne ground raid on the compound on the night of May 1–2, 2011.

As the SEALs from the JSOC based in Jalalabad, Afghanistan, penetrated Pakistani airspace in stealth versions of the Black Hawk helicopter and two larger Chinooks, RQ-170 drones flew overhead, monitoring the compound. When the helicopters arrived below, the drones provided high-resolution night-vision imagery of the SEALs landing at the compound and attacking. This imagery was sent to President Obama and national security team members, who were watching the events in Pakistan unfold on screen in the White House Situation Room. The stealth drones also used their eavesdropping equipment to monitor electronic transmissions among the Pakistanis, who had not been notified of the incursion.

Despite the odds against it, the deep raid into Pakistan was successful, and the SEALs managed to kill bin Laden and return to Afghanistan with his body.[114] Although the Pakistanis did launch fighter jets at the last minute to intercept the invaders, they were too late; the Americans had already retreated to Afghanistan in their stealth helicopters.

The Pakistani military and intelligence communities were humiliated when they realized that bin Laden had been living close to one of their bases (no evidence has emerged to show the bin Laden worked with the Pakistanis) and that they had been unable to protect their airspace from deep incursion. Relations between the Pakistanis and the Americans once again soured, and in a special session of the Pakistani parliament, ISI chief Ahmed Shuja Pasha denounced the American raid. The Pakistanis promised to reevaluate their relationship with the Americans and subsequently closed down the drone air base at Shamsi and ordered U.S. personnel to leave it. (American personnel returned to the base soon thereafter.) Pakistan's parliament also passed a resolution declaring that the drone strikes were a violation of sovereignty equivalent to the raid on bin Laden's compound in Abbottabad.[115]

But much good came out of the raid as well. The cache of information found in bin Laden's computers enabled the CIA to use a drone to track and kill Atiyah abd al Rahman, the man who was promoted to al Qaeda's number-two spot following bin Laden's death.[116] In fact there was a blitz of drone strikes in the FATA in the days and weeks after bin Laden's death as the CIA exploited information found in the slain terrorist's computers to strike at al Qaeda hideouts in the region.[117]

DRONES HELP HUNT DOWN GADDAFI IN LIBYA

American drones also played a major role in the 2011 air war against the Libyan dictator Muammar Gaddafi. In the winter and spring of that year Libyan citizens joined the so-called Arab Spring, a region-wide movement that was toppling Middle Eastern dictators, and began to openly resist Gaddafi's rule. In response, the dictator launched his troops on the rebels and promised to hunt them down like "rats." Alarmed by the prospect of a mass slaughter of civilians, President Obama, with the aid of the British and French, decided to wage an air campaign to assist the outgunned anti-Gaddafi rebels. U.S. and NATO aircraft began targeting Gaddafi positions to bolster the popular uprising and prevent a massacre. This was to be a unique airborne war that the Obama administration could claim did not "involve the presence of U.S. ground troops, U.S. casualties or a serious risk thereof."[118]

Not surprisingly, the United States made considerable use of drones in support of the insurgents. According to Global Security's John Pike, "The

US had neither the support nor the means to invade Libya. It would've been both a political and military blunder. So we had robots do the work for us— and it worked, perfectly. Qaddafi's air defenses and armor were obliterated from control rooms a world away."[119] The drones took out antiaircraft defenses, multiple rocket launchers, tanks, and truck-mounted weapons. In fact, twice as many drone strikes were carried out in the 2011 Libyan campaign (146) as in Pakistan that year.[120] More than 140 Hellfire missiles were fired by U.S. drones that were based in Sicily and flown from Creech Air Force Base in Nevada and elsewhere.[121]

The drones' most important strike occurred on October 20. On that day Gaddafi, who had been largely defeated and encircled by rebels in the town of Sirte, was spotted by high-flying drones. According to the *Telegraph*, the drones "built up a normal pattern of life picture so that when something unusual happened this morning such as a large group of vehicles gathering together, that came across as highly unusual activity and the decision was taken to follow them and prosecute an attack." As Gaddafi broke from cover and tried to escape in an eighty vehicle convoy, a drone spotted vehicles, tracked their movements, and relayed them to nearby French fighter jets. The drone then attacked the convoy with its missiles, bringing it to a halt whereupon it was attacked by the French jets. A bloodied Gaddafi survived the attack, but as he attempted to escape from the burning convoy, he was found in a nearby drainage ditch and executed by vengeful rebels.[122]

SENTINEL DOWN: THE CIA LOSES A TOP-SECRET SPY DRONE OVER IRAN

In 2009 a grainy photo of a mysterious jet drone taking off from the runway at a U.S. air base in Kandahar, Afghanistan, was leaked to the press. Technology wonks described the aircraft as the "Beast of Kandahar." Later the public learned that the top-secret stealth drone was the RQ-170 Sentinel, designed and built by Lockheed Martin's "Skunk Works," the company's supersecret division that built the U2 spy plane, F-22 jet fighter, and F-117 Nighthawk. Speculation about the RQ-170's purpose abounded. Why would the CIA deploy a radar-proof stealth drone to a region where the Taliban tribesmen had no radar or antiaircraft capabilities?

The first part of the answer came in May 2011, when the bat-wing Sentinel was used to spy on bin Laden's compound in Abbottabad. But that was not the only target of the stealth jet drone. On December 4, 2011, the Western media reported that the Iranian army's electronic warfare unit had downed a Sentinel that had intruded deep into Iranian airspace to spy on suspected nuclear weapons development facilities. The Sentinel, which had active phased-array radar, a "next-generation radar that allows one to use radar while staying stealthy," and full motion sensors to detect radiation, was said to be one of a fleet of drones that had been flying over Iran from a base in Shindand in western Afghanistan without the knowledge of the Afghan government.[123] The Iranians claimed the plane was captured intact over the eastern Iranian city of Kashmar, some 140 miles into Iran, by severing its communications links. The Iranians also claimed that they had reconfigured the drone's GPS coordinates to make it land in Iran, not Afghanistan. An Iranian engineer said, "The [Sentinel's] GPS navigation is the weakest point. By putting noise [jamming] on the communications, you force the bird into autopilot. This is where the bird loses its brain."[124]

Cyberwarfare experts in the West were, however, highly skeptical of the Iranians' claims of hijacking the state-of-the-art drone. According to an article in the *Christian Science Monitor*,

> To accomplish such a cyber coup, at least three and probably many more major technical hurdles would need to be overcome, several US cyber warfare and drone experts said. While none of these steps is impossible, each is difficult, and taken together, they represent a massive technological challenge for any enemy hacker—one that the US experts suggest is beyond Iran's capabilities.
>
> Hijacking and safe landing of the RQ-170 in Iran, if true, would represent a new level of cyber intrigue in the drone wars. First, Iran would need to spot the stealth drone. Second, it would need to jam the encrypted GPS signal. Third, it would have to substitute a false signal that the internal systems on the drone could understand and obey. US experts say even the first task—spotting the drone—would be very difficult for Iran.
>
> The US used a stealth helicopter in the Osama bin Laden raid that went undetected, and US military planes have bombed all over the

world undetected. So for Iran to detect a far-smaller stealth drone seems "almost like science fiction," says John Bumgarner, chief technology officer for the US Cyber Consequences Unit, a nonprofit cyberwarfare think tank.[125]

For its part, the Pentagon told reporters that there was no indication that the drone was brought down by "hostile activity of any kind."[126] To allay fears, the military pointed out that the drone was programmed to automatically erase sensitive data onboard to prevent it from falling into enemy hands.

Regardless of whether the CIA lost control of the drone on its own through a command malfunction or whether it was hijacked, the capture was a triumph for an Iran that was suffering under U.S.-led sanctions. Although the data onboard might have been erased or too encrypted to access, the Iranians were able to acquire the drone's top-secret, radar-proof "skin." The loss of the drone was such a disaster that the United States contemplated sending in airborne troops to retrieve it or blow it up but rejected the mission as too dangerous.[127] They feared that such a mission would be construed as an "act of war."

The Iranians subsequently displayed the stealth Sentinel in triumph in front of an American flag that had skulls in place of the stars.[128] A commander of Iran's Islamic Revolution Guards proudly proclaimed that the Sentinel, which was equipped with highly advanced surveillance systems, electronic communication, and radar systems, had fallen into the trap of Iran's armed forces.[129] When the U.S. government asked for the drone back, an Iranian official contemptuously said, "No one returns the symbol of aggression to the party that sought secret and vital intelligence related to the national security of a country."[130] In a further display of contempt, the Iranians offered to return the drone in the form of a 1:80-scale colorful toy replica of the drone, which began to sell in Iran in January 2012, complete with a stand featuring the words, "We will trample America under our feet."

In addition, Iran filed a complaint with the UN against the United States for the violation of its airspace. The Iranians promised that they would "reverse engineer" (i.e., duplicate) and "mass-produce" America's most secret drone. Although this seemed unlikely, the CIA feared Iran would share American secrets with other nations, most likely Russia and China, which were suspected to have the capacity to reverse engineer the drone. One U.S. expert bemoaned the loss of the drone by calling it "the biggest Christmas present to our enemies in probably a decade."[131]

THE SALALA FRIENDLY FIRE INCIDENT AND THE 2011–2012 DRONE LULL

According to the *Long War Journal*, there were 117 drone strikes in Pakistan in 2010 but only sixty-four in 2011.[132] This bucked the trend that had seen an increase in the number of drone strikes every year since 2005. There were four reasons for this unprecedented dip in the number of drone strikes. The first was the previously mentioned monthlong moratorium on drone strikes following the Pakistanis' arrest of the CIA contractor Raymond Davis in April and March 2011. This moratorium was designed to mollify the Pakistanis, who were infuriated by his killing of two Pakistani robbers/ISI agents. The second was a moratorium on drone strikes following the May 1–2 JSOC raid on bin Laden's compound in Abbottabad, which similarly infuriated the Pakistanis.

The third reason received less attention. In the summer of 2011 the Obama administration had an internal debate on the usefulness of signature killing. The catalyst for this debate was the clumsy Datta Khel strike, which killed numerous civilians (discussed previously). Discussion of this collateral damage tragedy among U.S. officials ultimately led to a tightening of restrictions on signature killings of low-ranked Taliban fighters.

The story of the tightened restrictions was broken in November 2011 by the *Wall Street Journal*, which reported that the CIA had backed down on its "wide net" policy of killing low-ranked Taliban foot soldiers after U.S. military and diplomatic officials complained that they were damaging America's tenuous relationship with Pakistan. Although they recognized the strikes' overall tactical effectiveness, members of the U.S. military and State Department had asked the CIA to be more selective in their targets. According to the *Wall Street Journal*, "The disputes over drones became so protracted that the White House launched a review over the summer, in which Mr. Obama intervened."[133]

In the end the CIA conceded to the more restrictive parameters for signature killings. One senior U.S. official described the new drone strike standards as follows: "The bar has been raised. Inside CIA, there is a recognition you need to be damn sure [the strike] is worth it."[134] In addition, the State Department won greater sway in influencing strike decisions, Pakistani leaders were to be given advance notice about future drone operations, and the CIA agreed to suspend drone strikes when Pakistani officials visited America. Soon thereafter the pace of drone strikes in Pakistan began to drop off.

The fourth reason for the decrease in drone strikes was the so-called Salala incident, on November 26, 2011, which led to an almost two month lull in attacks. The Salala incident was the lowest point in Pakistani-U.S. relations since Bush and Musharraf had forged the post-9/11 alliance. It involved a case of friendly fire on the Afghan-Pakistani border that led to the death of twenty-four Pakistani soldiers, including two officers, at the hands of American airpower.

The incident began on the night of November 25 and early morning of November 26 when a joint Afghan-U.S. patrol was fired upon from a position on the Pakistani side of the border. Believing that they were taking fire from Taliban militants, the U.S.-Afghan force called in air support from Apache Longbow attack helicopters, an AC-130 gunship, and at least one F-15. The U.S. air assets attacked a Pakistani army checkpoint on Salala Mountain in the Mohmand Agency, which was known by its code name "Volcano," where they thought the fire originated from. This checkpoint then radioed to a nearby checkpoint known as "Boulder," which fired antiaircraft weapons at the intruding helicopters.

The helicopters, however, returned to attack a second time as the Pakistanis frantically tried to contact U.S., Afghan, and NATO forces to call off the air assault. When communication was finally established, the attack was terminated, but not before twenty-four Pakistani soldiers at checkpoint "Volcano" had been killed and thirteen wounded. The primary fault lay in poor communication between the Pakistanis and the Afghan-U.S. force as well as a lack of information on their respective positions in an area where Taliban fighters were known to operate. Each side blamed the other.[135]

Regardless of who fired first, the killing of twenty-four Pakistani soldiers on Pakistani soil caused outrage across Pakistan. Having barely tolerated CIA spies, drone strikes, border raids, and political pressure to do more in the FATA for the years, not to mention the SEALs' ground attack against bin Laden, the Pakistani military and political establishment were furious. Obama personally apologized for the incident: "For the loss of life—and for the lack of proper coordination between U.S. and Pakistani forces that contributed to those losses—we express our deepest regret. We further express sincere condolences to the Pakistani people, to the Pakistani government, and most importantly, to the families of the Pakistani soldiers who were killed or wounded."[136] But his words were not enough.

As the bodies of the slain Pakistani army "martyrs" were publicly buried, there were anti-American protests throughout the country. A Pakistani official asked, "How can anyone expect a regime in Islamabad to be giving more support to the U.S. when our soldiers are being killed in cold blood?"[137]

The Pakistani response to the Salala incident came a day after the attack. First, they closed down the NATO supply lines that allowed convoys to transport supplies from the port of Karachi through the Khyber Agency to troops in Afghanistan. These supply routes were not reopened until July 2012, when Secretary of State Hillary Clinton apologized for the Salala incident. Second, and most importantly for the drone campaign, the Pakistanis gave the Americans fifteen days to vacate the premises of the "secret" drone base at Shamsi in Baluchistan. America moved to respond with alacrity, and U.S. cargo planes removed equipment and personnel from the base before the December 11 Pakistani deadline.[138] The Pakistanis also threatened, "Any object entering into our airspace, including U.S. drones, will be treated as hostile and be shot down."[139] At this time anti-Americanism rose to a fever pitch among Pakistani army officers who felt that the Salala incident was a purposeful action intended to punish Pakistan for not tackling the Haqqani Network in North Waziristan.

Thus did the historic drone war, which began with the night killing of Nek Muhammad on June 18, 2004, and which accelerated to more than a hundred strikes in 2010, come to a sudden and unexpected halt in November 2011. The last drone strike of the 2011 campaign was on the night of November 15–16, 2011, in South Waziristan, the very territory where the first strike on Muhammad was carried out more than seven years earlier.[140] From this point forward, a Pakistani official has said, the drone strikes will most likely be limited to "high profile targets."[141] Another Pakistani official said, "There is likely to be some arrangement on drone attacks, with Pakistan calling for large reductions in their number and geographic scope, and demanding prior notification and approval of every strike."[142]

The post-Salala lull gave the Taliban and al Qaeda a much-needed reprieve from years of constant aerial bombardment. On January 6, 2012, the *New York Times* published an article titled "Lull in US Drone Strikes Aids Militants in Pakistan," which reported, "A nearly two-month lull in American drone strikes in Pakistan has helped embolden Al Qaeda and several Pakistani militant factions to regroup, increase attacks against Pakistani security forces and threaten intensified strikes against allied forces in Afghanistan, American and Pakistani officials say." The article ended by quoting a member of the newly emboldened Haqqani Network who said of the drones, which were flying overhead but now no longer firing, "There are still drones, but there is no fear anymore."[143] A FATA-based Pakistani security official said, "The militants were quite happy with this lull and they were publicly operating in the region as they were no more worried for their lives."[144]

That impunity ended on January 10, 2012, when a drone struck in Miranshah, North Waziristan, killing "at least four militants." A local eyewitness reported of the strike, "It was an unusually big bang. Since it was extremely cold I didn't leave the house, but could see a house on fire."[145] Reuters reported, "The latest drone strike in Miranshah appears to demonstrate that if there was any kind of moratorium on such attacks, it has now been lifted."[146]

Pakistanis were stunned by the development, especially since their government had promised that the drone strikes would now be limited to HVTs. But it turned out that the hit was in fact on an HVT. The January 10 strike targeted Alsam Awan, a British-Pakistani al Qaeda operative believed to be planning attacks on the West. An American official described Awan as "a senior Al Qaeda external operations planner who was working on attacks against the West." He added, "His death reduces Al Qaeda's thinning bench of another operative devoted to plotting the death of innocent civilians."[147]

Two days later the CIA let it be known that it was continuing the policy of going after HVTs when it launched a drone attack on an SUV and a car in the militant-controlled Datta Khel region of North Waziristan. As the dust cleared, it began to appear as if someone important had indeed been killed in the drone strike: Hakimullah Mehsud, the head of the Pakistani Taliban and the man responsible for killing the CIA drone team at Camp Chapman. But the Taliban adamantly denied the reports of the leader's death, claiming they were a CIA ruse to reveal his location.[148] In fact, Mehsud had been falsely reported to have been killed in both 2009 and 2010. It seemed that the wily Taliban leader had once again cheated death.

Regardless, two weeks after the attack, a Pakistani security agent told Reuters that the first two strikes of 2012 had been a "joint operation" between Pakistan and the United States. This source suggested that the two countries had inaugurated a new level of understanding and that Pakistani "spotters" had been used to track the terrorists targeted in the January 10 and January 12 strikes. This source also said, "Our working relationship is a bit different from our political relationship. It's more productive." He then provided for the first time details of how the Pakistanis work with the CIA in targeting terrorists: "We run a network of human intelligence sources. Separately, we monitor their cell and satellite phones. Thirdly, we run joint monitoring operations with our U.S. and UK friends. . . . Al Qaeda is our top priority." This source further explained that "Pakistani and U.S. intelligence officers, using their own sources, hash out a joint 'priority of targets lists' in regular face-to-face meetings. . . . Once a target is identified and 'marked,' his network coordinates with drone operators on the U.S. side. He said the United States bases drones outside Kabul, likely at Bagram airfield about 25 miles north of the capital. From spotting to firing a missile 'hardly takes about two to three hours.'"[149]

This extraordinary account of the murky CIA-Pakistani relationship would seem to indicate that the Pakistanis and the Americans had buried the ax following the Salala incident and agreed to continue to work together against their common enemy. It also indicated that the drones were once again operational, and after an almost two-month lull, the CIA's hunt for Taliban and al Qaeda militants in the hills of Pakistan's FATA region was on once again. The winter and spring 2012 campaign was halted briefly for two weeks in April after the Pakistani parliament voted to end the strikes, but it continued apace soon thereafter despite repeated Pakistani condemnations.

The strikes picked up in the late spring of 2012, after a brief lull. This period saw the dramatic killing of the new number two in al Qaeda, Abu

Yahya al Libi. Libi, a charismatic propagandist who had mocked the Americans on many videotapes calling for jihad, oversaw the day-to-day running of al Qaeda's external operations while bin Laden and Zawahiri kept a low profile. Libi had actually been captured previously in Afghanistan, but he had escaped from jail in 2007. After surviving several close drone calls, Libi was finally killed early in the morning on June 4, 2012, in a strike on the village of Hassu Khel in North Waziristan. On hearing the welcome news of Libi's death, an American official stated, "Zawahri will be hard-pressed to find any one person who can readily step into Abu Yahya's shoes. In addition to his gravitas as a longstanding member of A.Q.'s leadership, Abu Yahya's religious credentials gave him the authority to issue fatwas [decrees], operational approvals and guidance to the core group in Pakistan and regional affiliates. There is no one who even comes close in terms of replacing the expertise A.Q. has just lost."[150]

The death of the skilled al Libi was only the latest killing in the drone war of attrition that had largely dismantled al Qaeda in Pakistan by the spring of 2012. But even as al Qaeda's core group was systematically taken out in hundreds of strikes in Pakistan, the organization's regional franchises began to emerge, especially in Yemen, posing a new threat to America.

DRONE STRIKES IN SOMALIA, YEMEN, AND THE PHILIPPINES

The spring 2012 drone campaign was not limited to Pakistan's tribal regions. Ten days after the failed attempt to kill Hakimullah Mehsud, a CIA drone struck and killed an al Qaeda terrorist in the northeast African country of Somalia. There a militant group known as al Shabab, which had publicly aligned itself with al Qaeda, was trying to take over the country and enforce strict sharia law. Al Shabab had carried out scores of suicide bombings in Somalia; sent a suicide bomber to Uganda, where he killed seventy-six soccer fans watching the World Cup, including an American; and dispatched terrorists to neighboring Kenya, where they killed scores. The group's interest in projecting terrorism beyond Somalia alarmed the CIA.

Many foreigners, including some Americans of Somali descent from Minneapolis, and Arabs had come to fight alongside al Shabab militiamen and to help them create a strict Islamic state in Somalia and export terrorism. Among them was Bilal al Berjawi, a British-Lebanese Arab who was number two in al Qaeda's Somali operations, next in rank after an operative who had directed the 1998 bombings of the U.S. embassies in Kenya and Tanzania. He was in charge of the al Shabab's overseas recruitment, training, and tactics and responsible for the Uganda bombing.

But Berjawi's career with al Qaeda and al Shabab, which began in 2006, did not last long into the new year. On January 22, 2012, three missiles fired from a drone hit his car near the Somali capital of Mogadishu. Al Shabab issued a statement on the death of this valued operative: "The martyr received what he wished for and what he went out for, as we consider of him and Allah knows him best, when, in the afternoon today, brother Bilal al-Berjawi was exposed to bombing in an outskirt of Mogadishu from a drone that is believed to be American. He was martyred immediately."[151] The UK's *Guardian* added a fascinating detail about Berjawi's assassination: "The 27-year-old's wife is understood to have given birth to a child in a London hospital a few hours before the missile strike, prompting suspicions among relatives that his location had been pinpointed as a result of a telephone conversation between the couple."[152] In all likelihood the drone that killed Berjawi was flown from a new CIA air base in either the nearby Camp Lemonier, Djibouti, or Arba Minch, Ethiopia (although one source claimed the actual drone was flown by JSOC).[153]

The strike that killed Berjawi was not the first drone strike in Somalia, nor was it the first JSOC operation in the country. Two al Qaeda leaders had previously been killed in a special operations raid and a bombing there.[154] The first reported drone strike in Somalia had occurred on June 23, 2011, and was aimed at two al Shabab/al Qaeda members linked to American-Yemeni al Qaeda leader Anwar al Awlaki. This strike, which was carried out by JSOC, did not kill its intended targets but wounded them. According to the *Washington Post*, the strike was aimed at a Shabab convoy near the group's southern base at Kismayo and might have killed a senior al Qaeda leader named Ibrahim al Afghani, who had not been on the original target list.[155] In September 2011 local Somalis reported that three Shabab targets were again hit in the area of Kismayo by either CIA or JSOC drones.[156]

Interestingly enough, these two reports of drone attacks in Somalia led the Iranian *Press TV* to publish an exaggerated report that fifty-six drone strikes in Somalia had supposedly killed a total of 1,370 people. Although the

Press TV reports were uncritically picked up and published as fact by other networks, they were either a flight of fantasy or Iranian propaganda aimed at inciting anti-Americanism.[157] Regardless, the strikes in Somalia increased the number of countries where the CIA or JSOC carried out drone operations to seven (the others being Yemen, Afghanistan, Iraq, Pakistan, the Philippines, and Libya). It is worth noting that the killer Reaper and Predator drones operating in Somalia have been aided by smaller observation drones, such as Ravens, Scan Eagles, and Fire Scout remote-controlled helicopters, as well as massive Global Hawks.[158] In the future unarmed drones may work in conjunction with killer drones to hunt and kill targets.

Following the lull in strikes after the Salala incident, there was an uptake in drone surveillance and strike operations in Yemen in the spring of 2012. These strikes were largely a response to AQAP's four attempts to blow up jetliners with hidden bombs (including the infamous underwear bomber incident) and successful assassination of almost a hundred Yemenis with a suicide bomb that spring. AQAP had become the most lethal al Qaeda threat following the destruction of most of al Qaeda Central's membership in the Pakistani drone campaign, and the CIA felt the need to respond.

To compound matters, the chaos in Yemen after President Ali Abdullah Saleh stepped down in February 2011 allowed al Qaeda militants and an allied group known as Ansar al Sharia to seize power in much of Yemen's coastal Abyan Province. Their ultimate aim was, like that of al Shabab in Somalia, to overthrow the country's secular government and to establish strict sharia law.

In response to these alarming developments, the Obama administration tasked the CIA and JSOC with ramping up an assassination campaign designed to kill AQAP terrorists and weaken related Ansar al Sharia militants. The Pentagon and Langley decided to turn to drones when Tomahawk cruise missile strikes proved to be too clumsy; on one occasion the cruise missiles had caused the deaths of dozens of bedouin civilians.[159]

The U.S. drone attacks had started off slowly in 2009 but reached a crescendo in the spring of 2012. The stepped up pace of the Yemeni drone campaign can be seen in the following statistics: there were four airstrikes in 2009, ten in 2011, and forty-two in 2012. According to the *Long War Journal*, which monitors the air campaign, 322 militants and eighty-two civilians have been killed as of March 2013.[160] The percentage of civilian deaths is higher in the Yemeni strikes than in the Pakistani drone campaign, but still the Yemini campaign has been the subject of less controversy and opposition than the Pakistani operations. This is largely because the new president of Yemen, Abdu Rabu Mansour Hadi, has worked closely with the United States to counter the terrorist threat to his country. The Yemeni air force, for example, has also bombed Ansar al Sharia targets in Abyan Province in an attempt to dislodge the militants.[161] The Arab media has reported that U.S. trainers are working directly with Yemeni forces to help them retake districts lost to AQAP.[162]

The Pentagon and CIA drones used in the Yemeni campaign appear to be based at either Camp Lemonier, Djibouti (home to Combined Joint Task Force–Horn of Africa) or at an undisclosed base in Saudi Arabia. The majority of the drone strikes appear to target Shabwah Province, a known hideout for AQAP, which has the support of local tribes there, and Abyan Province, which was largely taken over by Ansar al Sharia militants in 2011. Yemeni sources have reported that JSOC and CIA drones are directly assisting the Yemeni military operations on the ground in these areas.[163]

The drone strikes in Yemen and the more widely publicized strikes in Pakistan show similar trends. For example, in Yemen the drones seem to be targeting vehicles that can be easily tracked and monitored, instead of houses, in order to avoid civilian collateral damage deaths. As in Pakistan, civilians have nonetheless been killed in Yemen, and this has led to protests by angry relatives of the slain civilians. The United States has also expanded its campaign in Yemen from more limited personality strikes to signature strikes, as happened in Pakistan in 2008.

The uncanny precision of the strikes also indicates that the CIA has established a network of spies and informers in Yemen that has been relaying the positions of terrorists to drones, as in Pakistan's FATA. Some members of Yemen's parliament have protested the drone strikes, and militants have started brutally killing those who are said to be spies working to help guide the drones; both of these trends were also found in the Pakistani tribal areas.

As occurred in the early days of the drone strikes in Pakistan, the Yemeni government has tried to deflect domestic criticism of the CIA strikes by claiming that their own air force carried them out. According to cables published on Wikileaks, former Yemeni president Saleh told Gen. David

Petraeus in 2010, "We'll continue saying the bombs are ours, not yours," and Deputy Prime Minister Rashad al-Alimi even admitted to lying to the Yemeni parliament about the American role in the drone strikes.[164] Those who knew Yemen's air force capabilities saw through such claims; the outdated Yemeni MiG 23s and 29s were incapable of making precision strikes on moving vehicles, especially at night.

The major differences between the Yemeni and Pakistani campaigns have been the more prominent role of JSOC in Yemen, the more direct role Obama plays in choosing the targets, and the large role that the drones played in supporting the Yemeni army in its ground operations against the militants in the spring of 2012. The drones, for example, were used to blow up AQAP ammunition depots and to hit the militants' defensive positions.

The most notable drone assassination strike in Yemen was the May 6, 2012, killing of Fahd al Quoso, an al Qaeda operative involved in both the AQAP plot to bomb passenger planes and the 2000 bombing of the USS *Cole* in Yemen. Quoso, who had $5 million bounty on his head, was killed in a personality strike after a Saudi double agent pretending to be an al Qaeda bombing volunteer relayed his coordinates to the CIA. The killing of Quoso as he exited a vehicle was in every way a double-agent reversal of the previous killing of the CIA team in Camp Chapman, Afghanistan, by an al Qaeda triple agent.

In addition to the campaigns in Somalia and Yemen, in February 2012 news of a drone strike on the remote Muslim island of Jolo, in the Philippines, began to surface. The strike killed fifteen militants who belonged to the pro–al Qaeda groups Abu Sayyaf and Jemaah Islamiyah.[165] Among those killed were three of the Philippines's most-wanted terrorists, Zulkifli bin Hir, Gumbahali Jumdail, and Mumanda Ali. These terrorists were involved in the 2002 Bali bombing, which killed or maimed hundreds of Indonesians and foreign tourists (mostly Australians), and in kidnapping to fund terrorism. According to the Washington, D.C.–based Jamestown Foundation, "The airstrike was reported to have been 'U.S.-led' and facilitated by an unmanned U.S. drone which tracked down Jumdail by honing in on a sensor that was placed by local villagers pretending to be seeking medical assistance from Jumdail."[166] A previous drone strike in 2006 against a notorious Filipino terrorist named Umar Patek had been claimed by the Philippine government to deflect criticism.[167]

The strikes in Somalia, Yemen, and the Philippines demonstrate how drones are increasingly used in areas similar to the FATA where a weak or nonexistent central government is combating lightly armed paramilitaries, terrorists, or insurgents who have only rudimentary air defenses. Although it is all but impossible to carry out "snatch operations" in these hostile tribal lands, the remote-control drones give the U.S. military and CIA unprecedented capability to reach out and kill terrorists or insurgents in areas they would otherwise be unable to access.

REPORTS THAT DRONES ATTACK CIVILIANS AT FUNERALS AND KILL RESCUERS WHO COME TO AID DRONE VICTIMS

There has been no group more active in lobbying against the drones than the London-based Bureau of Investigative Journalism. In February 2012 the bureau published a damning article that was picked up by the *New York Times*, London's *Sunday Times*, and other newspapers and bloggers across the globe. The article, titled "Obama Terror Drones: CIA Tactics in Pakistan Include Targeting Rescuers and Funerals," reported that the drones had killed numerous civilians while they were either trying to help those previously targeted by drones or attending funerals for drone victims.[168] A *New York Times* account of the bureau's investigation, titled "US Said to Target Rescuers at Drone Sites," read as follows:

> British and Pakistani journalists said Sunday that the C.I.A.'s drone strikes on suspected militants in Pakistan have repeatedly targeted rescuers who responded to the scene of a strike, as well as mourners at subsequent funerals. The report, by the London-based Bureau of Investigative Journalism, found that at least 50 civilians had been killed in follow-up strikes after they rushed to help those hit by a drone-fired missile. The bureau counted more than 20 other civilians killed in strikes on funerals. The findings were published on the bureau's Web site and in The Sunday Times of London.[169]

After reading the report, antidrone activist Clive Stafford Smith, a lawyer who heads a British-American charity called Reprieve, which is opposed to drones, said that such "double-tap" strikes "are like attacking the

Red Cross on the battlefield. It's not legitimate to attack anyone who is not a combatant." Christof Heyns, the UN special rapporteur on extrajudicial executions as of 2010, concurred with Stafford Smith and added, "Allegations of repeat strikes coming back after half an hour when medical personnel are on the ground are very worrying. To target civilians would be crimes of war."[170] Such reactions were typical among readers, who doubtless envisioned the drones firing on responding paramedics and concerned civilians desperately trying to dig fellow civilians out of the rubble of a drone strike.

What most stories in the media that covered the Bureau of Investigative Journalism's investigation did not report were the further details found on the bureau's website. According to the site,

> The first confirmed attack on rescuers took place in North Waziristan on May 16, 2009. According to Mushtaq Yusufzai, a local journalist, Taliban militants had gathered in the village of Khaisor. After praying at the local mosque, *they were preparing to cross the nearby border into Afghanistan to launch an attack on US forces. But the US struck first* [author's emphasis].
>
> A CIA drone fired its missiles into the Taliban group, killing at least a dozen people. Villagers joined surviving Taliban as they tried to retrieve the dead and injured. But as rescuers clambered through the demolished house the drones struck again. Two missiles slammed into the rubble, killing many more. At least 29 people died in total.
>
> "We lost very trained and sincere friends," a local Taliban commander told The News, a Pakistani newspaper. "Some of them were very senior Taliban commanders and had taken part in successful actions in Afghanistan. Bodies of most of them were beyond recognition."[171]

Essentially, the civilians killed in this drone strike were assisting Taliban militants who, before the drone struck them, had been "preparing to cross the nearby border into Afghanistan to launch an attack on US forces." The civilians were thus aiding and abetting active Taliban militants whom the Pakistani, Afghan, and U.S. governments consider terrorists. The U.S. government had previously dropped leaflets in the FATA warning local tribesmen that if they assisted the terrorists, they would share their fate.[172]

Three independent studies discussed in chapter 7 have demonstrated that approximately 95 percent of those who are targeted in drone strikes are militants. Thus it follows that the vast majority of those who are being removed or rescued from the rubble of a drone strike are Taliban militants or al Qaeda terrorists, not civilians. In fact, in many, if not most, cases those who are removing the victims from the rubble are themselves Taliban militants; there are very few if any emergency medical technicians, paramedics, or first responders in this undeveloped area. The Taliban militants are the de facto authorities in these regions, so their presence at the scenes of attacks is not surprising.

There are scores of media reports of the Taliban "cordoning off" drone strike zones and "conducting recovery operations."[173] A typical account reads, "A local resident said he was woken by two loud explosions around 4 a.m. on Thursday. Militants rushed to the site immediately after the attack to clear the rubble and retrieve the bodies, he said, speaking on the condition of anonymity."[174] A second report states, "First a volley of four missiles hit a compound in the village of Mizar Madakhel. After Taliban fighters cordoned the area and began to recover bodies, a second volley was fired. Initial reports indicated that 12 Taliban fighters were killed."[175] A third source reads, "Eight militants were killed and two wounded. Militants have surrounded the [targeted] compound and are removing the dead bodies."[176] Another local Pashtun source claims, "The reason why these estimates about civilian 'casualties' in the US and Pakistani media are wrong is that after every attack the terrorists cordon off the area and no one, including the local villagers, is allowed to come even near the targeted place. The militants themselves collect the bodies, bury the dead and then issue the statement that all of them were innocent civilians."[177] A BBC story similarly reported, "Officials say that local Taliban militants immediately cordoned off the [strike] area and closed the road in the aftermath of the attack."[178] The Aryana Institute for Regional Research and Advocacy, a think tank of researchers and political activists from the North-West Frontier Province and FATA, similarly reported, "People told me that typically what happens after every drone attack is that the Taliban and Al-Qaeda terrorists cordon off the area. No one from the local population is allowed to access the site, even if there are local people killed or injured."[179] Civilians are rarely able to rush to the scene of a drone strike on Taliban terrorists and insurgents in order to help wounded militants or retrieve their bodies.

So well-known is the Taliban's propensity to cordon off areas where their comrades have been killed or wounded in a drone strike that a FATA-based Pakistani official even offered the Americans some advice on how to kill more Taliban using drones. According to Al Jazeera, "He explained that after a strike, the terrorists seal off the area to collect the bodies; in the first 10–24 hours after an attack, the only people in the area are terrorists. You should hit them again—there are no innocents there at that time."[180] FATA-based scholar Farhat Taj similarly wrote, "Your new drone attack strategy is brilliant, i.e., one attack closely followed by another. After the first attack the terrorists cordon off the area and none but the terrorists are allowed on the spot. Another attack at that point kills so many of them. Excellent! Keep it up!"[181]

Clearly the CIA has taken this advice and, on the basis of many reports of Taliban militants rushing to the scenes of drone attacks to save their buried or wounded comrades, begun targeting those who arrive at drone strike locations to rescue wounded militants. The Bureau of Investigative Journalism surely realized that in the vast majority of cases those who are killed or wounded in the drone strikes are themselves Taliban militants and that those who are killed in follow-up strikes are more than likely also Taliban militants. Yet they chose to completely omit this important detail in their scathing report.

As for the claim that the drones target funerals, we have already seen in chapter 1 that the CIA fired on a funeral for slain Taliban commander Sangeen Khan in 2009 in an effort to kill Pakistan's most-wanted man, Baitullah Mehsud, who was in attendance. A local source claimed that of the sixty-seven people killed in that drone explosion eighteen were villagers.[182] Clearly the CIA felt that on this occasion the risk of killing civilian bystanders at a Taliban-organized funeral for a militant was outweighed by the opportunity to kill the terrorist mastermind who had sent scores of suicide bombers into Pakistani towns killing and maiming hundreds if not thousands. Ironically, among the Taliban suicide bombers most-favored targets have been funerals of tribal chieftains who have resisted them.[183]

There has been one other famous drone strike on a Taliban funeral. Taj describes it as follows:

On the other hand, drone attacks have never targeted the civilian population except, they informed, in one case when the funeral procession of Khwazh Wali, a TTP commander, was hit. In that attack too, many TTP militants were killed including Bilal (the TTP commander of Zangara area) and two Arab members of al Qaeda. But some civilians were also killed. After the attack people got the excuse of not attending the funeral of slain TTP militants or offering them food, which they used to do out of compulsion in order to put themselves in the TTP's good books. "It [this drone attack] was a blessing in disguise," several people commented.[184]

Even though the majority of rescuers at drone strikes on Taliban militants are themselves Taliban militants and the rare strikes on funerals have been aimed at notorious terrorists, the bureau's skewed reporting created a popular image of drones pouncing on concerned "first responders" and innocent civilians mourning their dead at funerals. This public relations fiasco certainly helped paint a distorted image of the drones as operating beyond the pale of humanity.

9

The Argument for Drones

We've seen violent extremists pushed out of their sanctuaries.
We've struck major blows against al Qaeda leadership as well as the
Taliban's. They are hunkered down. They're worried about their
own safety. It's harder for them to move, it's harder for them to train
and to plot and to attack . . . and all of that makes America safer.
—President Barack Obama

The people of Waziristan are suffering a brutal kind of
occupation under the Taliban and al Qaeda. It is in this context
that they would welcome anyone, Americans, Israelis, Indians
or even the devil, to rid them of the Taliban and al Qaeda.
Therefore, they welcome the drone attacks.
—Farhat Taj, Pashtun scholar from the FATA

I n a rare commentary on the CIA's Predator/Reaper drone campaign in Pakistan, in May 2009 former CIA director Leon Panetta said, "Very frankly, it's the only game in town in terms of confronting or trying to disrupt the al Qaeda leadership."[1] Those who advocate for the aerial assassination campaign agree with Panetta and offer a simple, unassailable argument recognizing its benefits: it is killing large numbers of Taliban and al Qaeda leaders and foot soldiers, disrupting their military and terrorist operations, and sowing fear and dissension among the enemy. This saves civilian lives because it is hard for the terrorists to plan mass-casualty attacks when they themselves are being terrorized. The strikes are the ultimate form of deterrence and are saving countless civilians from future terrorist attacks against the West, Pakistan, and Afghanistan. One U.S. official called the drone strikes "the purest form of self-defense."[2] An example of this is the drone strike on June 3, 2011, that preemptively killed Ilyas Kashmiri, a Pakistani terrorist mastermind who had been assigned the task of carrying out an assassination attempt on President Obama.[3] In conventional military terms, the attack on Kashmiri could be described as "suppression fire," which is meant to kill the enemy or keep him pinned down and thus unable to fight.

The drone campaign advocates argue that those in the West who are against the drones are naive and have selective memories. They have forgotten, or deliberately overlooked, the hundreds upon hundreds of suicide bombings in Pakistan and Afghanistan that have slain or maimed scores and the horrors of 9/11 and the 7/7 bombings in London, which were carried out by al Qaeda–linked militants trained in the FATA or Afghanistan's tribal lands. Antidrone activists also seem to live in an alternative universe where talk of bona fide terrorists who have been targeted and killed by drones simply does not exist. Instead there is a total focus on unintentional civilian casualties that result from strikes on these unmentioned terrorists. Had the FATA-trained Faisal Shahzad successfully set off his bomb in Times Square or had the FATA-based Rashid Rauf blown up numerous passenger jets with liquid bombs, many of the antidrone voices in the West would be muted, if not silent, it can be argued.

Those in Pakistan who are against the drones forget that the Taliban have deliberately killed thousands of their compatriots on a yearly basis. The drones are the front line in the defense of Pakistani civilians, who are threatened by terrorists living in a de facto Taliban terrorist state in the FATA.

Perhaps the best example of the way the drones have saved civilian lives is the case of the previously mentioned Mumbai-style terrorism plot in Europe that was disrupted by drones. As the FATA-based terrorists plotted to use bombs and automatic weapons to slaughter civilians in France, Germany, and Britain, they themselves were hunted down and killed by drones, and thus countless civilian lives were spared. Grateful British security officials subsequently downgraded their terrorism threat level and said,

"Strikes have decimated the Al Qaeda senior leadership, and we didn't have to get directly involved."[4]

Similarly, in my own work in Afghanistan in 2009 with the Afghan National Directorate of Security, I discovered that most suicide bombers in Afghanistan (the world's second largest recipient of suicide bombings at the time) were trained in madrassas and terrorist camps in the FATA.[5] Having been trained to be "Mullah Omar's missiles," the suicide bombers were sent into Afghanistan to detonate their explosives and slaughter Afghan civilians. The Afghan police and intelligence officers I worked with all applauded the drones for disrupting potential Taliban terrorist plots and killing future suicide bombers and terrorists before they could make their way to Afghanistan to wreak havoc on civilians.

Another example of the deterrent effect of drones is the case of the Taliban leader Qari Hussain. Qari Hussain was known as the *Ustad e Fedayeen* (Teacher of Suicide Bombers). He ran a camp in South Waziristan that trained suicide bombers who then went into Pakistan and slaughtered hundreds of innocent civilians. Pakistani journalist Syed Saleem Shahzad wrote of Hussain,

> He moved back to South Waziristan and soon won notoriety for brutally killing anti-Taliban figures and for introducing the practice of slitting the throats of Pakistani soldiers. . . .
>
> He established a reign of terror across the [Swat] valley that had once been known for its tranquility, beauty and peace-loving residents. One of his more gruesome habits was to teach valley militants how to slit a throat with a rusty knife, film the incident and then distribute it on a video recording. By now the small-fry sectarian agitator had evolved into a national terror ringmaster.[6]

One of Hussain's typical suicide bombing attacks targeted a group of Pashtun elders who were meeting in the FATA to muster a *lashkar* (militia) to fight the Taliban in Orakzai Agency. A Taliban suicide bomber broke into the meeting and set off a bomb that slaughtered approximately a hundred. In one of his more horrific acts, Qari Hussain kidnapped children and brainwashed them so that they would blow themselves up as suicide bombers.

On more than one occasion the Pakistani security forces discovered and closed down his "suicide nurseries." After raiding one of his child indoctrination schools, a Pakistani officer said,

> It was like a factory that had been recruiting nine- to 12-year-old boys, and turning them into suicide bombers. The computers, other equipment and literature seized from the place give graphic details of the training process in this so-called "nursery." There are videos of young boys carrying out executions, a classroom where 10- to 12-year olds are sitting in formations, with white band of Quranic verses wrapped around their forehead, and there are training videos to show how improvised explosive devices are made and detonated.[7]

Qari Hussain also trained Faisal Shahzad, the Pakistani American who tried to set off a car bomb in a civilian-packed Times Square.

In response to such outrages the Pakistani government put an approximately $600,000 bounty on Hussain's head and began a hunt for him in South Waziristan. In desperation the Pakistanis asked the Americans for help, and the CIA made several attempts to kill Hussain with a drone. As the U.S. drones hunted Qari Hussain and his fellow suicide-bombing trainers, a Pakistani wrote a letter to his country's primary English-language newspaper, *Dawn*, in support of the drone strikes. He also pointed out the hypocrisy of Pakistanis who reflexively protested the U.S. drone strikes for accidentally causing collateral damage deaths in their hunt for terrorists but who did not protest against suicide bombers for purposefully killing civilians. This author wrote,

> When American pilotless aircraft, the drones, zero in on and attack the masterminds of these suicide attacks, in the tribal area, the religio-political parties raise a storm of protest on the grounds that the sovereignty of Pakistan has been threatened. The media too, inadvertently, follows the line of the religio-political parties and creates a hype and makes it look as if the Americans have done great harm to Pakistan while the other set of foreigners, i.e. Arab, Chechen, Uzbek militants, have played no role in a persistent effort to destabilize Pakistan.

Probably, the media and, in turn, the general public forget that the vast majority of the militant leaders that plan suicide attacks inside Pakistan are the former students of the seminaries controlled by the very leaders who are in the forefront to raise a storm of protests when an isolated drone attack takes place by the Americans. But these leaders observe absolute silence when the militants carry out suicide attacks that inflict devastating damage on Pakistan's human and material assets.

Seen neutrally, it will dawn on critics of the drone attacks that the Americans are assisting Pakistan by annihilating the masterminds that sit in the tribal areas, plan, prepare and dispatch suicide attackers who play havoc with life and property in the urban Pakistan.[8]

Another person, who supported the drone strikes for the same reason, posted a comment on *Dawn*'s website: "Innocent women and children are also dying in our neighborhoods, kindergarten schools and in our shopping malls in suicide attacks. I don't think they deserve to die either. I guess drone attacks are good as long as they [are] killing those terrorists and terrorist sympathizers. Maybe you will understand this when somebody from your neighborhood dies in a suicide attack."[9]

A third Pakistani wrote in *Dawn* similarly condemning the suicide bombers while condoning the drones:

Why would the people under Taliban and Al Qaeda occupation and oppression not cheer when these murderers are killed? What does not make sense is the chorus of protests over these drone attacks emanating from people like Imran Khan and Hamid Gul—to name only two—who claim to speak for the people of the tribal areas. What exactly is their agenda, and why are they acting as cheerleaders for these terrorists?

I have often wondered about this callous hypocrisy too. If we condemn the Americans so vociferously over the drone campaign, should we not be more critical of the thugs who are killing far more Pakistani civilians? And yet, it seems that our more popular Urdu anchorpersons and TV chat show guests reserve their outrage for Washington, while giving the Taliban and Al Qaeda a free pass over their vicious suicide bombings that have taken hundreds of innocent lives in recent weeks.

Why then are we silent over the daily killings of fellow Pakistanis by the TTP and other terror groups, while frothing at the mouth over the drone attacks? Clearly, this irrational and double-faced reaction is based in the anti-American sentiment that has taken root in Pakistan.[10]

As the drones hunted Qari Hussain, some in Pakistan saw the CIA's remote-control killers as an ally in the struggle against the scourge of suicide-bombing terrorism. After the 2009 killing of Baitullah Mehsud, Pakistani president Ali Zardari declared, "Due to his death the Taliban leadership is in disarray, the major suicide bombing network and Taliban patronage has been disrupted. Acts of terror have considerably decreased in the border area."[11] This sort of acknowledgment of the drones' usefulness may have increased when it was reported that a CIA drone tracked and blew up a Taliban suicide-bombing truck packed with explosives before it could reach its intended target.[12]

As the debate continued, on October 4, 2010, the drones finally found Qari Hussain and fired on his vehicle. Hussain was injured in the strike, and three of his guards were killed. Three days later the Teacher of Suicide Bombers himself was finally killed in a second drone strike. As they had after Baitullah Mehsud was killed, many Pakistanis quietly celebrated the death of the man who had killed so many of their people with his suicide bombers. The Pakistani *Express Tribune* published the comments of readers who wrote in to celebrate the CIA drone assassination of Hussain. Their commentary demonstrates that not everyone in Pakistan was loudly opposed to the drone strikes on the Taliban; some clearly supported them. Some of the readers' comments follow:

If this news is true then God Bless America and God Bless the drones. He blew people into pieces and he got blown into bits and pieces. Good riddance!
    —A Suhail

Wonderful news, if confirmed. This guy had the blood of thousands of innocent Muslims and Pakistanis on his hand. Good riddance to bad rubbish. Drone attacks may be wrong but they are effective and precise compared to full-fledged army onslaught which normally has more collateral damage.

—Hamood

I wish the drones could pick these guys up so that they could be hanged in public.

—QB 2

Yet another triumph of allied forces in the war on terror. Much of the elite leadership of these terrorists has either been killed or captured as a result of joined efforts by the allied forces.

—Yasir Qadeer

If true, this would be an achievement on the scale of eliminating Baitullah Mehsud. This Qari Mehsud was the "Ustad-e-Fedayeen," the master trainer of suicide bombers. Unfortunately this cockroach has been reported killed before, most recently in January, only to pop up later in a video and resume his threats and vile ways.

—Bangash

Bravo! Drone attacks will have to continue if we want Pakistan to be free of these monster Jihadis.

—Adam[13]

The successful drone strike on Qari Hussain took out a key Taliban terrorist whose primary mission was to kill civilians with suicide bombings and is as positive a testimony for the drone campaign as any. The killing of Hussain may have saved the lives of hundreds, if not thousands, of Pakistanis. It also reaffirmed the message that the CIA had been sending to the militants: that they had gone from being the hunters to the hunted.

The Israelis, who also have a drone program and a long history of hunting their enemies for assassination, have also found that their killing campaigns have had a profound disruptive effect on terrorist activities. According to Daniel Byman,

Israel's experience shows that a sustained campaign of targeted killings can disrupt a militant group tremendously, as slain leaders are replaced by less experienced and less skilled colleagues. This can lead the group to make operational and strategic mistakes, and over time, pose less of a danger. Moreover, constant killings can create command rivalries and confusion. Most important, the attacks force an enemy to concentrate on defense rather than offense. To avoid becoming targets, group leaders must minimize communications, avoid large groups, constantly change their locations, disperse their cells, and take other steps that make it far harder for them to do the sustained, systematic planning required to build large organizations and carry out sophisticated attacks.[14]

Did the drone strikes have this effect on the Taliban and al Qaeda in the FATA? The answer seems to be a resounding yes. Whereas the Taliban previously gathered in large numbers to demonstrate their power, hold rallies, plot, train, equip, administer harsh shariah justice, and publicly enforce their writ in the FATA, they can no longer do so thanks to the threat of drones. One FATA resident said, "The Taliban will never gather in large number in broad daylight to be targeted by the drones."[15] David Rohde, the American journalist held captive by the Haqqani Network in North Waziristan, provided the following eyewitness account of the drones' disruptive effect on terrorist operations:

> During my time in the tribal areas, it was clear that drone strikes disrupted militant operations. Taliban commanders frequently changed vehicles and moved with few bodyguards to mask their identities. Afghan, Pakistani, and foreign Taliban avoided gathering in large numbers. The training of suicide bombers and roadside bomb makers was carried out in small groups to avoid detection. . . .
>
> The drones were terrifying. From the ground, it is impossible to determine who or what they are tracking as they circle overhead. The buzz of a distant propeller is a constant reminder of imminent death. Drones fire missiles that travel faster than the speed of sound. A drone's victim never hears the missile that kills him.[16]

The drones' potential targets have to operate under the assumption that the unseen killers are watching them at all times with an unblinking eye. A writer for *Esquire* magazine who was embedded with a drone crew in Afghanistan reported,

> The insurgents and terrorists hate the drones; that is certainly true. U.S. soldiers drive around Iraq and Afghanistan always waiting, wondering if this will be the moment they blow up. So it is for insurgents. Missiles launched from UAVs are America's version of the roadside bomb, infecting insurgents with the same paranoia and fear. "He knows we're there. And when we're not there, he thinks we might be there," Colonel Theodore Osowski, who commanded the unit in Kandahar, told me. "It's kind of like having God overhead. And lightning comes down in the form of a Hellfire."[17]

The mere possibility of a strike helps disrupt or prevent terrorist and insurgent activities. The Taliban and al Qaeda cannot, for example, do simple tasks such as drive in SUV convoys or communicate on cell phones, much less transport weapons and openly train without fear of being tracked and killed by unseen killers in the sky. One Taliban commander provided insight into the considerable lengths the militants in the FATA go to just to avoid being detected and killed by the hovering drones:

- If a drone is heard, fighters must disperse into small groups of no more than four people.
- Satellite or SMS [short message service, a form of text messaging on mobile phones] forms of communication are no longer used. All communications are done orally or by code.
- Meetings are announced only at the last minute, with nothing planned in advance in order to avoid leaks. Even senior commanders do not know the precise location of regional commanders.
- Taliban leaders have reduced the size of their security escorts to one or two men "in whom they have complete confidence."[18]

Such security precautions certainly curtailed the Taliban and al Qaeda's ability to function. Maintaining security precautions in and of itself becomes a time-consuming distraction. A *Washington Post* article based on interviews in the FATA provides further insight into how the drones have terrorized the terrorists: "Militants who once freely roamed markets and helped settle disputes, tribesmen said, have now receded to compounds. Fighters shun funerals and trackable technology. They rely on motorbikes or their feet to move, pro-Taliban tribesmen said. Insurgent leaders, the highest-value drone targets, move 'three times in a night,' said Malik, the Pakistan army commander. 'The militants are desperate,' said a Miranshah teacher, 38, adding that residents pray that drones hit their targets."[19]

One local official said of the Taliban, "Their freedom of movement has been curtailed to a great extent. This has caused demoralization. There is no discrimination while taking out targets, be they Pakistani Taliban or Al Qaeda and their foreign affiliates."[20] A Pashtun tribal elder reported, "I know that the Taliban and Al Qaeda are very much in trouble because of these attacks. These drone attacks really strike fear in them."[21] The *Guardian* similarly reported, "In Wana, the capital of south Waziristan, foreign fighters are shunning the bazaars and shops, and locals are shunning the fighters. 'Before, the common people used to sit with the militants,' said Wazir. 'Now they are also afraid.'"[22]

As reports like this indicate, foreign al Qaeda fighters in particular—but also rank-and-file Taliban members—have become magnets for drone strikes, and the locals are less willing to protect them or give them sanctuary in their homes. It is difficult for the Taliban and al Qaeda to plot new terrorist outrages against Pakistan, Afghanistan, and the West when, as former CIA chief Michael Hayden put it, their sanctuary in the FATA is "neither safe nor a haven."[23] An article in the *New York Times* vividly captured the disruptive impact the drones have had on the militants' former safe haven:

> Tactics used just a year ago to avoid the drones could not be relied on, [a member of the Taliban] said. It is, for instance, no longer feasible to sleep under the trees as a way of avoiding the drones. "We can't lead a jungle existence for 24 hours every day," he said.
>
> Militants now sneak into villages two at a time to sleep, he said. Some homeowners were refusing to rent space to Arabs, who are

associated with Al Qaeda, for fear of their families' being killed by the drones, he said. The militants have abandoned all-terrain vehicles in favor of humdrum public transportation, one of the government supporters said.

The Arabs, who have always preferred to keep at a distance from the locals, have now gone further underground, resorting to hide-outs in tunnels dug into the mountainside in the Datta Khel area adjacent to Miram Shah, he said.

"Definitely Haqqani is under a lot of pressure," the militant said. "He has lost commanders, a brother and other family members."[24]

According to one source, Haqqani and his followers, the most active terrorists in eastern Afghanistan, can no longer communicate with al Qaeda by cell phone owing to the threat of drone strikes.[25] The terrorists, especially the foreign al Qaeda element, also bemoan the local Pashtun unwillingness to associate with militants who have become lightning rods for drone strikes. Local Pashtun tribesmen also stay away from Taliban militants, who are known to be targets for drone strikes, and this has made the Taliban's sanctuary less secure. A militant claimed that the Taliban were constantly on the alert for the sound of approaching drones and said, "We now often sleep in the river beds or under the eucalyptus trees."[26]

One al Qaeda message lamented the impact of the strikes: "The harm is alarming, the matter is very grave. So many brave commanders have been snatched away by the hands of the enemies. So many homes have been leveled with their people inside them by planes that are unheard, unseen and unknown."[27] This war of attrition has hurt al Qaeda's ability to plot. According to Jane Mayer,

Surviving militants are forced to operate far more cautiously, which diverts their energy from planning new attacks. And there is evidence that the drone strikes, which depend on local informants for targeting information, have caused debilitating suspicion and discord within the ranks. Four Europeans who were captured last December after trying to join Al Qaeda in Pakistan described a life of constant fear and distrust among the militants, whose obsession with drone strikes had led them to communicate only with elaborate secrecy and to leave their squalid hideouts only at night.[28]

Many Taliban leaders, fearing local spies, have fled from the tribal areas to cities in non-Pashtun urban areas, such as Quetta in Baluchistan or the southern port town of Karachi, seeking a safer sanctuary. This has put them farther from the field of operations in Afghanistan and the FATA.[29] A Saudi terrorist named Najam, for example, lost his legs in the June 2012 drone strike that killed Abu Yahya al Libi and fifteen other militants and was forced to flee Pakistan for his native Saudi Arabia to recuperate.[30] Al Qaeda members have also attempted to flee FATA for Yemen to escape the drones.

Noticing this trend, a senior American counterterrorism official said, "The enemy is really, really struggling. These attacks have produced the broadest, deepest and most rapid reduction in al-Qaida senior leadership that we've seen in several years."[31] Former CIA director Hayden said, "A significant fraction of al-Qaeda senior leadership in the tribal region has been 'taken off the battlefield.'"[32] Former CIA chief Panetta similarly claimed, "These operations are seriously disrupting al Qaeda. . . . It's pretty clear from all the intelligence we are getting that they are having a very difficult time putting together any kind of command and control, that they are scrambling and that we really do have them on the run."[33]

Some Pakistanis seemed to agree with this assessment, and a senior Pakistani intelligence official estimated, "Some 60 to 70 percent of the core al Qaeda leadership has been eliminated, dealing a serious blow to the network's capacity to launch any major attacks on the West."[34] Taliban prisoners have told their interrogators that the drone attacks have taken a psychological toll as well. One source has them saying, "Hey, we're doing all the dying out here. How much longer can we put up with this?"[35] In 2011 an al Qaeda leader admitted, "There were many areas where we once had freedom, but now they have been lost. We are the ones that are losing people, we are the ones facing shortages of resources. Our land is shrinking and drones are flying in the sky."[36]

As a result of the attacks, insecurity and distrust grew among al Qaeda members. According to a U.S. official, "They have started hunting down people who they think are responsible. People are showing up dead or disappearing."[37] A second counterterrorism official said, "This last year [2008] has been a very hard year for them. They're losing a bunch of their better leaders. But more importantly, at this point they're wondering who's next."[38]

The constant threat of attack or surveillance has forced the Taliban and al Qaeda to dismantle their training camps in favor of hidden classrooms or dugouts in the mountains. One Taliban commander said, "Arab nationals and key al-Qaeda members never stayed together and often spent the nights underground or in caves."[39] A Taliban commander in North Waziristan similarly explained, "In the early days of our jihad, our training camps were visible and people would come and go. We were not so concerned about the security of our locations, but that is all changed now. We abandoned all our old camps and re-located to new places."[40] A Taliban militant added, "We don't even sit together to chat anymore."[41] Al Qaeda members in particular have also given up on using cell phones as a means of communication for fear that they will be tracked and killed, as Nek Muhammad and countless other terrorists were. The same holds true for Yemen. A Yemeni who was opposed to AQAP and related militants said, "Al Qaeda hates the drones, they're absolutely terrified of the drones . . . and that's why we need them."[42]

The deaths of so many high-level al Qaeda leaders has also meant that many mid-level operatives who are inexperienced and lacking direct ties to bin Laden have been elevated to higher positions in the organization. In fact at least three number threes in al Qaeda and three number twos have been killed by drones. Dozens of lower-ranked al Qaeda leaders have also been killed. With each strike their replacements are becoming less skilled and experienced. Intelligence officials report that as a result, the caliber of al Qaeda's plots has "degraded"; they are now "strikingly amateurish compared with the attacks of Sept. 11, 2001, and other airline plots that followed."[43]

According to the *Japan Times*, the former number three in al Qaeda complained to bin Laden that the drones were killing operatives faster than they could be replaced.[44] The *Washington Post* pointed out that prior to his death, bin Laden had received numerous e-mails from his followers complaining about the toll taken by CIA drones.[45] Former White House counterterrorism chief John Brennan summed up his predictions for the drone strikes' effects on al Qaeda's ranks saying, "If we hit al-Qaida hard enough and often enough there will come a time when they simply can no longer replenish their ranks with the skilled leaders they need to sustain their operations."[46]

In summary, then, perhaps the best argument for the drone strikes is that they work strategically as a means to break down the Taliban and al Qaeda's ability to kill civilians in the West and in Afghanistan and Pakistan. Drone strikes also act as a critical means of tactical deterrence and have been responsible for killing hundreds, if not thousands, of Taliban and al Qaeda leaders and rank-and-file members. Those who have not been killed have lost their ability to take advantage of their safe haven in the FATA to openly plot and carry out new mass-casualty and insurgent attacks in the region.

The *New York Times* best summed up the effects of the drone strike on the enemy: "With their ranks thinned by a relentless barrage of drone strikes, some experts believe Al Qaeda's operatives in Pakistan resemble a driver holding a steering wheel that is no longer attached to a car."[47] The fact that many of the cars that Taliban terrorists drive in Afghanistan and Pakistan are vehicle-borne improvised explosive devices (VBIEDs) used to deliberately kill civilians might be the ultimate rationale for continuing the strikes.

DRONES ARE THE SAFEST, MOST HUMANE WAY TO KILL TERRORISTS

Drones are indisputably the most humane, precise, and clean way for the American officials tasked with defending America from al Qaeda terrorists (who have already killed thousands of civilians) to protect the United States, the West, Afghanistan, and Pakistan from future mass-casualty terrorism. This statement is especially true when one considers potential alternative methods, such as arrests (which are not an option because the Pakistanis have forbidden U.S. ground forces on their land) or aerial bombardments by clumsy, conventional manned aircraft of the sort used in bombing campaigns in Afghanistan, Iraq, Kosovo, and Bosnia.

The significant decline in U.S.-Pakistani relations following the September 2009 raid on Musa Nika and the May 2011 Navy SEAL incursion into Pakistan to kill bin Laden also indicate how fraught with political implications unilateral, ground-force operations to kill and capture terrorists are in the sovereign nation of Pakistan. The same certainly holds true for Somalia and Yemen. Even if the Pakistanis allowed U.S. special force raids, the 1993 Black Hawk Down incident in Mogadishu, Somalia, serves as a potent reminder of how disastrous snatch operations can be in human terms. In that incident, eighteen U.S. servicemen were killed and seventy-three wounded while they tried to arrest an enemy warlord in territory held by hostile insurgents.

Remote-control drones, in contrast, do not put our soldiers' or pilots' lives at risk, and this is an important distinction given that America has already lost more than six thousand men and women to the wars in Iraq and Afghanistan. In military terms, the drones "project capability without projecting vulnerability." But most importantly, drones also keep civilian deaths to a minimum, as described in detail in chapter 7. The *Wall Street Journal* best summed up this argument for relying on the drones:

> The case is easy. Not even the critics deny its success against terrorists. Able to go where American soldiers can't, the Predator and Reaper have since 9/11 killed more than half of the 20 most wanted al Qaeda suspects, the Uzbek, Yemeni and Pakistani heads of allied groups and hundreds of militants. Most of those hits were in the last four years. . . . The civilian toll is relatively low, especially if compared with previous conflicts. Never before in the history of air warfare have we been able to distinguish as well between combatants and civilians as we can with drones. Even if al Qaeda doesn't issue uniforms, the remote pilots can carefully identify targets, and then use Hellfire missiles that cause far less damage than older bombs or missiles. Smarter weapons like the Predator make for a more moral campaign. . . .
>
> International law also allows states to kill their enemies in a conflict, and to operate in "neutral" countries if the hosts allow bombing on their territory. Pakistan and Yemen have both given their permission to the U.S., albeit quietly. Even if they hadn't, the U.S. would be justified in attacking enemy sanctuaries there as a matter of self-defense.[48]

Another *Wall Street Journal* article, titled "Predators and Civilians: An Intelligence Report Shows How Effective Drone Attacks Are," similarly makes the case for the drones as the most discriminate and humane way to fight the terrorists:

The argument against drones rests on the belief that the attacks cause wide-scale casualties among noncombatants, thereby embittering local populations and losing hearts and minds. If you glean your information from wire reports—which depend on stringers who are rarely eyewitnesses—the argument seems almost plausible.

Yet anyone familiar with Predator technology knows how misleading those reports can be. Unlike fighter jets or cruise missiles, Predators can loiter over their targets for more than 20 hours, take photos in which men, women and children can be clearly distinguished (burqas can be visible from 20,000 feet) and deliver laser-guided munitions with low explosive yields. This minimizes the risks of the "collateral damage" that often comes from 500-pound bombs. Far from being "beyond the pale," drones have made war-fighting more humane.[49]

Dan Goure, vice president of the Lexington Institute, similarly argued that drones are "more precise and careful than other alternatives." He pointed out that the drones' capacity to loiter and wait for the right moment to kill their target when there are no civilians around makes them "much more discriminate and therefore moral, ethical, and effective than what we had before."[50] William Saletan made a similar argument in an article titled "In Defense of Drones: They're the Worst Form of War, Except for All the Others":

> How do drones measure up? Three organizations have tracked their performance in Pakistan. Since 2006, Long War Journal says the drones have killed 150 civilians, compared to some 2,500 members of al-Qaida or the Taliban. That's a civilian casualty rate of 6 percent. From 2010 to 2012, LWJ counts 48 civilian and about 1,500 Taliban/al-Qaida fatalities. That's a rate of 3 percent. Drones are like laparoscopic surgery: They minimize the entry wound and the risk of infection. . . . Over the years, I've shared many worries about the rise of drones. But civilian casualties? That's not an argument against drones. It's the best thing about them.[51]

Democrat Senator Dianne Feinstein supported this rationale for deploying drones when she said, "What this does is it takes a lot of Americans out of harm's way . . . without having to send in a special ops team or drop a 500-pound bomb."[52] Journalist Noah Shachtman similarly argued, "Taken together, it might be the most precise, most sophisticated system for applying lethal force ever developed—the Platonic ideal of how an air war should be run."[53] Georgetown scholar Christine Fair adds,

> If we know little about the drone strikes, we know enough about the alternative means of eliminating terrorists in FATA to know that they're probably worse. Pakistan has no police in FATA to arrest them. The Pakistan army is now in its 13th month of sustained combat in the region, an effort that has flattened communities and displaced millions but done little to chip away at the insurgents' strength. Drone strikes may not be perfect, but they're likely the most humane option available.[54]

Similarly, the *Economist* opined,

> Civil-liberty advocacy groups have raised concerns about targeted killings by drones of suspected terrorists, especially in the case of al-Awlaki, who was an American citizen. But so far, the use of drones has not fundamentally challenged the Geneva Convention–based Law of Armed Conflict. This requires that before an attack, any weapons system (whether manned or unmanned) must be able to verify that targets are legitimate military ones, take all reasonable precautions to minimise civilian harm and avoid disproportionate collateral damage.
>
> As long as a UAS pilot can trust the data from remote sensors and networked information, he or she should be able to make a proper assessment based on the rules of engagement in the same way as the pilot of a manned aircraft. Indeed, because of the unique characteristics of UAS, he may be in a better position to do so. He should have more time to assess the situation accurately, will not be exhausted by the physical battering of flying a jet and will be less affected by the adrenalin rush of combat.[55]

In other words, drones take away much of the pressure of high-speed combat flying and allow pilots more time to reflect on life-and-death decisions.

But not only Westerners support this perspective. The president of Yemen, Abdu Rabu Mansour Hadi, has compared the advanced drones hunting al Qaeda to his country's clumsy MiG fighter-bombers. After saying that he personally approves of every U.S. drone strike in his country, he has added, "They pinpoint the target and have zero margin for error," and their "electronic brain's precision is unmatched."[56] An article in the *Globe and Mail* featured an interview with a Pakistani ISI official who also supported the drone strikes on the basis of their precision and ability to avoid the sort of large numbers of dead civilians and destruction that all too often stemmed from full-scale Pakistani military operations and aerial bombardments against the Taliban. The paper reported, "Not everybody worries about the drones. Asad Munir, former station chief in Peshawar for Inter-Services Intelligence, Pakistan's leading spy agency, is among many observers who argue that precision strikes cause vastly less damage than the alternative of sending ground forces into Waziristan. Soldiers kill civilians with stray artillery or bombs, he said, whereas the drones only make mistakes when they're given the wrong intelligence. 'They don't make more enemies,' he said."[57] A Pakistani civilian agreed: "Drones strike with a lot of precision they mostly kill the terrorists, and we are happy with that. It's our army that fires indiscriminately."[58]

Then there is the problem of population displacement that comes from full-scale antiterrorist operations carried out by the Pakistani military as an alternative to pinpoint drone strikes. As the Taliban execute civilians, terrorize populations, burn schools, enforce shariah law contrary to Pakistani laws, dispatch suicide bombers to kill and maim, and attack Pakistani police and military in their jihad against the Pakistani state, the Pakistani army has to respond. The Pakistani army's resulting clumsy operations against the Taliban in Swat, Bajaur, South Waziristan, and elsewhere have destroyed houses and whole communities with artillery and aerial bombardments. This has forced millions of Pakistanis to flee these zones for their lives. According to John Schmidt, author of *The Unraveling: Pakistan in the Age of Jihad*, Pakistani military operations against the Taliban in the appropriately named Operation Earthquake in South Waziristan led to the destruction of four thousand homes and the displacement of 200,000 people. Another five thousand homes were destroyed in military operations in Bajaur that displaced 300,000 people. A further 3 million Pakistanis were made refugees by Pakistani military operations in the Swat Valley.[59] A Pakistani author from the FATA region summed up the benefits of the drones compared to these full-scale Pakistani army and air force operations:

> The people feel comfortable with the drones because of their precision and targeted strikes. People usually appreciate drone attacks when they compare it with the Pakistan Army's attacks, which always result in collateral damage. Especially the people of Waziristan have been terrified by the use of long-range artillery and air strikes of the Pakistan Army and Air Force. People complain that not a single TTP or al Qaeda member has been killed so far by the Pakistan Army, whereas a lot of collateral damage has taken place. Thousands of houses have been destroyed and hundreds of innocent civilians have been killed by the Pakistan Army.
>
> On the other hand, drone attacks have never targeted the civilian population except, they informed, in one case when the funeral procession of Khwazh Wali, a TTP commander, was hit.[60]

A survey among Pashtuns from the region found that in contrast to Pakistani army operations, "the drone attacks cause a minimum loss of innocent civilians and their property. The respondents appreciated the precision of such attacks."[61] The BBC similarly reported, "Zahid, from Wana, in South Waziristan, said: 'I think the drone strikes are good: they target the right people, the terrorists.' Bahar Wazir, from Shawal, North Waziristan, said: 'I prefer the drone attacks to army ground operations, because in the operations we get killed and the [Pakistani] army doesn't respect the honour of our men or women.'"[62] Another Pashtun said, "I have heard people particularly appreciating the precision of drone strikes. People say that when a drone would hover over the skies, they wouldn't be disturbed and would carry on their usual business because they would be sure that it does not target the civilians, but the same people would run for shelter when a Pakistani jet would appear in the skies because of its indiscriminate firing."[63]

An article for the Pakistani *Daily Times* titled "In Favor of Drone Attacks" reported, "According to the Aryana Institute of [*sic*] Regional

Research and Advocacy, 80 percent of tribals think that drones hit exact targets as pointed out on the basis of authentic information. They compare this with military operations that prove more destructive. In military operations, hundreds of homes are demolished, people are compelled to flee and civilian casualties become a natural thing. They also provide the Taliban with anti-army sentiment."[64]

Another interview with Pashtuns from the FATA found,

> The reasons why people living in the tribal areas might support drone strikes are rarely heard on Pakistani television. But Safi says they include both dislike of militants and fear of what alternative counterterror strategies entail. . . .
>
> All the [Pakistani army] ground operations have caused massive population displacements as people fled the conflict areas. Many refugees said they not only feared being caught in the crossfire but also becoming targets of retribution killings by either militants or troops as territory changed hands. . . . Safi says that he and many others in the tribal areas regard the drones as accurate and much less likely to kill civilians than is ground fighting.[65]

A Pashtun from the region writing for the *Daily Times* supported the drone strikes: "What is clear enough is that the drone strikes, however unpopular they may be, are likely to be more popular than the realistic alternatives: the Taliban's violence or the Pakistani army's operations, which have displaced millions."[66]

As for the Pakistani leadership, they reportedly appreciate the drone strikes, for they save them the Pakistani soldiers' lives that would surely be lost in fighting against the well-armed militants.[67] This is of paramount importance to the Pakistanis, who have lost more soldiers fighting the Taliban than the U.S.-led coalition has lost in Afghanistan. The United States has pressured the Pakistani government to invade North Waziristan (the focus of most drone strikes) to fight the Haqqani Network, but the Pakistanis are already stretched thin from fighting in South Waziristan, Swat Valley, and Bajaur. The drone campaign provides pinpoint killing of terrorists in North Waziristan and saves the Pakistanis from having to launch a full-scale, bloody invasion of this territory, which they consider to be controlled by "good Taliban" (i.e., Taliban fighters who are not at war with the Pakistani state). For that the region's inhabitants are undoubtedly grateful.

DRONES DON'T VIOLATE PAKISTANI SOVEREIGNTY; AL QAEDA AND THE TALIBAN DO

In October 2011 Pakistani writer Sayeda Asrar Bukhari summed up the feelings of many Pakistanis when he wrote, "Every time America launches a drone attack on our soil, it violates our sovereignty. By using its fight against terrorism as an excuse, America has killed thousands of our innocent citizens in the tribal areas."[68] Many average Pakistanis similarly speak in almost reflexive fashion about drones violating their sovereignty. But, as has been demonstrated in the previous chapters, both their elected leaders (Musharraf, Zardari, and Gilani) and their military leaders have actively supported the drone campaign—so much so that they have allowed the CIA to run drone strikes on the Taliban and al Qaeda from the Shamsi Air Base in Pakistan. If the United States is, or was, allowed to operate on Pakistani soil with Pakistani troops guarding the drone base at Shamsi, their operations cannot be termed a violation of sovereignty. The same certainly cannot be said for al Qaeda or the Taliban, which have openly declared a bloody war on Pakistan and have carved off much of the tribal lands from that state.

Yet many Pakistanis seem to be in a state of denial; they do not want to recognize that the Taliban or foreign al Qaeda fighters are a threat to their nation's ability to rule its own territory. They see Taliban militants as misunderstood fellow Pakistani Muslims who have been scapegoated by the "imperialist American infidels." They believe that the drones, not the terrorists, cause the terrorism. The truth, however, is that terrorism and militancy in Pakistan long predated the drone war of 2008. As the 2007 Lal Masjid incident, in which militants tried to take the capital hostage and enforce strict shariah law nationally, clearly demonstrated, the Taliban are very much the enemy of the Pakistani state. More than anything else, the bloody Lal Masjid episode drove the Pakistani Taliban to break its temporary and advantageous "truce" with the Pakistani government and declare a secessionist-terrorist jihad against it. Since then the Taliban and its al Qaeda allies have commenced a suicide bombing war/insurgency on Pakistan that has led to thousands of deaths and have conquered territory within a hundred miles of the Pakistani capital. The Taliban have also enforced its will in places as far away as relatively cosmopolitan Lahore, in the eastern Pakistani province of Punjab, where they forced vendors to burn DVDs and music and demanded the closure of an ancient red light district. These terrorist efforts have nothing to do with the drones or the CIA.

John Schmidt masterfully captured the urgency of the Taliban threat to Pakistan in his recent book:

> As the Taliban began to draw nearer to Punjab ordinary Pakistanis began to feel threatened. What had been a distant abstraction now loomed on the horizon as something very real. That was not the way most Pakistanis, the majority of them Barelvi followers of [moderate] Sufi Islam wished to be governed. The Pakistani Taliban were a threat to their way of life. . . .
>
> The barbarians were literally at the gates. The Pakistani political establishment, civilians and military, had been humiliated. Their policy of concession had been shown to be bankrupt. It suddenly seemed possible that the Pakistani Taliban would bring all the Pashtun lands west of the Indus under their sway.[69]

In the face of such onslaughts, Pakistani prime minister Yousuf Gilani told reporters in Lahore, "Pakistan is not fighting the war of any other country. The war on terror is in our own interests."[70] Noting this trend, *Washington Post* columnist David Ignatius wrote,

> The Pakistanis have their own heavy score to settle with the Taliban, whose bomb attacks have stretched from Peshawar to Lahore. The Pakistani spy service, the Inter-Services Intelligence directorate, has been a special target, with attacks on some of its senior officers and regional headquarters. That's one reason Pakistanis have been cooperative; they're angry and they want revenge.
>
> "It became personal for the ISI," said the senior administration official. Enraged by attacks on their colleagues, Pakistani officers have worked closely with the CIA to gather intelligence in tribal areas. The

Predator assault "has given the Pakistanis some breathing room," the administration official said.[71]

As the internal Taliban menace has become officially recognized by the Pakistani government and military, some in Pakistan have stopped their "violations of sovereignty" boilerplate rhetoric and begun to criticize the Taliban and al Qaeda for threatening their country's sovereignty. These voices in support of the drone strikes complain that the foreign fighters (Central Asian Uzbeks, Arabs, North Africans, Europeans, etc.) and Taliban militants (including Taliban fighters from Afghanistan) represent the greatest threat to Pakistan's sovereignty, not their U.S. allies. For example, Syed Alam Meshud, a Peshawar-based political activist who is from Waziristan, said, "To those people sitting in the drawing rooms of Islamabad talking about the sovereignty of Pakistan, we say, 'What about when Arabs or Uzbeks occupy your village? What about sovereignty then? We compare the drones with *Ababeel*—the swallows tasked by God in the Koran to smite an army with rocks. Any weapon which kills these people who damaged my sovereignty is in fact helping the sovereignty of my region."[72] In the same vein, Bashir Ahmad Gwakh, writing for Radio Free Europe, opined,

The fact that Al-Qaeda leaders (including Osama bin Laden, who was living in the garrison city of Abbottabad just a couple of hours' drive from Islamabad), foreign fighters, and Haqqani-led Afghan Taliban all live in Pakistan damages Pakistani credibility when it asks that drone attacks be stopped in respect of the country's sovereignty. If American drone strikes violate Pakistani sovereignty, what about all of the foreign militants who not only launch attacks across the border into Afghanistan but are also a huge security threat to the people of Pakistan?[73]

One Pakistani writing for the *Daily Times* gave the same argument:

Sovereignty is the complete power to govern a country. . . . Whatever little control Pakistan's establishment had there [FATA] is now being put an end to by the TTP, al Qaeda and its offshoots. The Taliban are

openly roaming around in FATA, alleged criminals are publicly executed in shariah courts, people are amputated and frequent attacks are
being carried out against our army. The demolition of schools has
become an old story. . . . The drones are targeting all those people who
are bent upon the real violation of our sovereignty and who are busy in
a declared war against our army and state machinery.[74]

Farhat Taj carried out a survey among Pashtuns from the FATA for the
Jamestown Foundation and found that

> the majority of the respondents (13 of 15) did not fully see the drone
> attacks as a violation of the sovereignty of Pakistan. Their argument is
> very simple: the state of Pakistan has already surrendered FATA to the
> militants. Therefore Pakistan has no reason to object to the drone
> attacks. Pakistan will have this right only if it can retake the areas from
> the militants. Some respondents said that their homeland is used by
> the militants and the ISI as a launching pad for attacks on ISAF and
> NATO forces in Afghanistan.[75]

The Pakistani paper the *News* published an account of a similar survey
in the FATA:

> Pakistan's sovereignty, they [the interviewees] argued, was insulted and
> annihilated by Al-Qaeda and the Taliban whose territory FATA is, after
> Pakistan lost it to them. The US is violating the sovereignty of the
> Taliban and Al-Qaeda, not of Pakistan. Almost half the people said
> that the US drones attacking Islamabad or Lahore will be violation of
> the sovereignty of Pakistan, because these areas are not taken over by
> the Taliban and Al Qaeda.
>
> Over two-thirds of the people viewed Al-Qaeda and the Taliban as
> enemy number one, and wanted the Pakistani army to clear the area of
> the militants. A little under two-thirds want the Americans to continue
> the drone attack because the Pakistani army is unable or unwilling to
> retake the territory from the Taliban.[76]

In the article "Why I Support Drone Attacks" on Chowk.com, Raza Habib echoes this sentiment: "Over the years the Pakistani establishment and a series of governments have literally watched helplessly as militants use those safe sanctuaries to promote terrorism in the mainland. If anything, the actual violation of sovereignty is being carried out by the militants rather than the drones which are aimed at eliminating them! Realistically speaking drones are helping the Pakistani state to establish sovereignty."[77]

Finally, an article published in the *Daily Times* titled "In Favor of Drone Attacks" argued,

> Civil society in Pakistan is well aware of the fact that the drones are cleansing us of terrorists but sometimes they fail to resist the temptation to speak out against them. This is caused by the extreme right trumpeting warnings of US encroachments on our sovereignty.
>
> First, Pakistan is the US's frontline non-NATO ally in the war against terror, which means that Pakistan and the US have to willingly extend their best possible support to each other. Drone attacks are now to be considered as committed support in that process. Second, the drones are targeting all those people who are bent upon the real violation of our sovereignty and who are busy in a declared war against our army and state machinery. They are not the ones who, if scared, will respect our sovereignty.
>
> What is our chief security threat? It is terrorism. What should we prioritize? No doubt its elimination. And what are drones doing, except the same? Rather than theoretical terms we need to think in terms of ground reality. We are faced with a severe threat and cannot channel our meager resources to military operations against the militants.
>
> The present government should not get blackmailed by rightist propaganda; it should boldly and publicly acknowledge the agreement with the US [regarding drone attacks] if there is any.[78]

Thus the conventional wisdom in the West that Pakistanis unanimously see the drone strikes as a violation of their country's sovereignty does not seem to hold true. A number of FATA natives, as well as Pakistanis elsewhere, see the drone strikes as the best antidote against the terrorists who have truly violated their country's sovereignty by carving parts of it off into off-limits, fundamentalist shariah law terrorist states. The same certainly holds true for the Pakistani military establishment and government, which have taken much criticism for their backing of the drone strikes but continue to tacitly support them despite their unpopularity.

PASHTUNS (AND OTHER PAKISTANIS) WHO FAVOR THE DRONE CAMPAIGN

Conventional wisdom among antidrone activists in the West is that the drones "make more enemies than they kill" and drive the FATA's tribemen to actively support the Taliban. There are, however, anecdotal accounts, such as that of Pervez Hoodbhoy, that describe a people who are seething as they suffer from the tyranny of the Taliban. Hoodboy writes,

> Many FATA students in my university have seen the barbarity of Taliban militants from close quarters. They want the beasts killed— and they don't care how and by whom. For example, a physics PhD student from Mohmand told me that he has not been back to his village for 3 years and still lives in constant fear of being kidnapped by militants. His crime? To have protested the public decapitation by the Taliban of 14 members of a neighbour's family outside the village mosque.
>
> Not surprisingly, Kurram's Shiite community of about half a million people is also said to be largely supportive of drone strikes. They have suffered an estimated 2,000 deaths at the hands of Taliban militants since 2007. Photographs of severed heads and limbs have been posted on the internet by the Taliban, who think that Shiites deserve nothing less.
>
> A scientific survey of attitudes in FATA in today's dangerous circumstances is impossible. Nevertheless, the impression one gets in talking to individuals is that tribal people with education generally favour drone strikes. This includes those who have lost relatives. But uneducated people, who form the overwhelming majority, hate them. . . .
>
> But the Taliban want something immensely more dreadful. They stone women to death, force girl-children into burqa, cut off limbs, kill

doctors for administering polio shots, threaten beard-shaving barbers with death, blow up girls schools, and kill musicians. In a society policed by Taliban vice-and-virtue squads, art, drama, and cultural expressions would disappear. The only education would be that of madrassas.[79]

Similarly, Aamir Latif reported,

Many in tribal areas remain hostile to the Taliban. Hazar says that the Taliban view tribal elders and prayer leaders as their main rivals and keep a close eye on them because they are generally respected by the civilians in the region. "We are better Muslims than the Taliban. We don't need their advice. We have already been following a decent way of life. What else do they want from us?" says Kamal Shah, a lawyer who heads the anti-Taliban *jirga*.

Hundreds of tribesmen displaced by the ongoing pitched battles between Pakistani security forces and Taliban militants in the restive northern tribal belt staged an anti-Taliban rally on Monday. This was the first time that the displaced tribesmen demonstrated against the Taliban, dubbing them responsible for their woes.[80]

Many other Pashtuns would seem to agree with these sentiments. Several studies demonstrate that there is some Pashtun support for the drone strikes against the region's grim Taliban masters. In 2008 the Pakistan-based Aryana Institute for Regional Research and Advocacy carried out a survey among tribesmen in the agencies of North Waziristan, South Waziristan, and Kurram that led to some surprising conclusions. Following are some of the survey questions and their responses:

- Do the militant organizations get damaged due to drone attacks? (Yes 60%, No 40%)
- Do you think the drones are accurate in their strikes? (Yes 52%, No 48%)
- Do you think anti-American feelings in the area increased due to drone attacks recently? (Yes 42%, No 58%)

- Should Pakistan military carry out targeted strikes at the militant organisations? (Yes 70%, No 30%)[81]

In other words, a majority of those questioned said the drones weaken the militants, are accurate, and don't lead to anti-Americanism. The study further found that

> the popular notion outside the Pakhtun [Pashtun] belt that a large majority of the local population supports the Taliban movement lacks substance. The notion that anti-Americanism in the region has increased due to drone attacks is rejected. The study supports the notion that a large majority of the people in the Pakhtun belt wants to be incorporated with the state and wants to integrate with the rest of the world. . . .
>
> The people I asked about civilian causalities in the drone attacks said most of the attacks had hit their targets, which include Arab, Chechen, Uzbek and Tajik terrorists of Al-Qaeda, Pakistani Taliban (Pakhtun and Punjabis) and training camps of the terrorists.[82]

These voices in favor of the drones were not an anomaly in the FATA. In December 2009 a coalition of FATA-based political parties and civil organizations opposed to terrorism issued the "Peshawar Declaration." Among other provisions, it stated,

- The conference demands that targeted and immediate operations against all centers and networks of terrorism should be initiated.
- This conference also demands the elimination of all foreign, non-local and local terrorists in FATA.[83]

The declaration also dealt with the drone attacks in detail:

> The issue of Drone attacks is the most important one. If the people of the war-affected areas are satisfied with any counter-militancy strategy, it is the Drone attacks which they support the most. According to the people of Waziristan, Drones have never killed any civilian. Even

some people in Waziristan compare Drones with Ababels (The holy swallows sent by God to avenge Abraham, the intended conqueror of the Khana Kaaba). A component of the Pakistani media, some retired generals, a few journalists/analysts and pro-Taliban political parties never tire in their baseless propaganda against Drone attacks.[84]

There is also a peace movement in the FATA known as Amn Teherek (aka Amn Tehrik) (the Peace Movement). Its platform has been described as follows: "The Amn Tehrek publicly opposes Taliban and Al-Qaida, denounces the 'strategic depth' madness, demands the military to conduct targeted operations against the militants in FATA and supports drone attacks. The Amn Teherek has expressed such views almost every month in its socio-political activism. Mainstream Pakistani media largely ignore them."[85] This group has established the Pakhtunkhwa Peace Forum, which had a Facebook page with such messages as "Drone proved its worth in FATA by killing the worst enemies of Pashtun and humanity. . . . Bravo Mr. drone you are the only source which helped us salvate from terrorists in region."[86]

When the CIA commenced a lull in drone strikes in late 2011, a group of Pakistanis launched an online petition calling for the Americans to restart their campaign in order to "save the lives of thousands."[87] Similarly a former Pakistani intelligence agent said, "Most of the population is a sort of hostage, and they know that these people have made our life miserable. They say— if there's a correct target—good riddance."[88] A 2010 Deutsche Presse Agentur article on the drone strikes offers further evidence that some people in the FATA region who are terrorized by the Taliban and al Qaeda support the drone strikes:

> The cooperation from reluctant Pakistani intelligence agencies might be due to constantly increasing pressure from Washington, but many residents in Pakistan's tribal region have come to see the drones as a blessing.
>
> "These drones give us a sense of protection—that there is some-one who is doing something against these people who kill innocent people in the name of Islam," said a resident of Miranshah, the main

town in North Waziristan, who asked to be identified as Shin Gul. Gul, 29, fears Taliban persecution if his real name was known. His brother was murdered two months ago when his father refused to marry his young daughter to a Taliban fighter.

"People in the tribal region have varying opinions on the drone attacks," said Nasir Dawar, a North Waziristan journalist who has covered dozens of the strikes. "Some people think they are doing some good, and some believe they are killing innocent people and challenge the Pashtu national honour," he said.

Dawar said he was convinced that the drone aircraft were mainly targeting the militants and that most of the civilians killed in the attacks were either from the extended families of the militants or victims of collateral damage. "I have never seen a missed hit," Dawar said, adding that the strikes were creating panic and fear among the militants. Once used to moving freely, senior Taliban and al-Qaeda leaders are now being forced to spend their nights in sleeping bags under a tree in the fields or in a mountain cave and hold emergency meetings in a moving vehicle instead of a building, Dawar said.[89]

A recent report by the Jamestown Foundation based on interviews in the FATA, titled "Drone Attacks: Pakistan's Policy and the Tribesmen's Perspective," supported these conclusions. After interviewing students from the region, the report concluded,

They agree that the government of Pakistan has no writ whatsoever over the tribal agencies. They hold the militant occupation responsible for:

- Damaging their culture and traditions.
- Eliminating their entire traditional and indigenous leadership.
- Weakening the tribal society.
- Occupying their houses by force.
- Destroying their traditional and democratic institution of jirga (an assembly of elders that makes decisions based on consensus) and tribal code of Pashtunwali ("The Way of the Pashtuns"), instead replacing it with the militants' own strict brand of Shari'ah.

■ Bringing destruction to homes and businesses by inciting Pakistani military operations[90]

Regarding the drones, the Jamestown Foundation pollsters discovered that the interviewees thought the following:

■ The drone attacks are killing the leadership of those al-Qaeda and other militant groups who have made ordinary tribesmen and women hostages. Ordinary people are powerless against the militants and drones are seen as helpful by eliminating the militants and frustrating the designs of ISI.

■ The drone attacks have resulted in substantial damage to the militants, especially the elimination of the Arab and Punjabi leadership of al-Qaeda.

■ The drone attacks cause a minimum loss of innocent civilians and their property. The respondents appreciated the precision of such attacks.[91]

In addition the article stated,

The respondents expressed a strong desire for drones as a means to attack the leadership of the local Pashtun Taliban. Half of those who supported drone attacks said that people's daily lives are affected most by the local Taliban and not the Arabs or other al-Qaeda militants who generally mind their own business, or have perhaps assigned the duty of harassment to the local Taliban. One of the respondents suggested that if only ten people amongst the leadership of the local Taliban were killed, the hierarchy of the organization would collapse like a house of cards.[92]

Christine Fair, an American scholar who visited the tribal zone, gave an account that backs up the Jamestown Foundation findings:

These FATA residents are strong proponents of the drones. They report that the drones are so precise that the local non-militants do not fear them when they hear the drones above as they are confident that they will hit their target. Locals attribute this precision in part to the placement of "targeting chips" which direct the ordnance to the exact location of the militants in their redoubts. The accurate placement of these chips requires local cooperation to provide the whereabouts of these militants. This has driven an important wedge between the locals and militants with the former shunning the latter.[93]

As these findings demonstrate, the notion that many in the West and the settled parts of Pakistan have of the FATA region being inhabited by pro-Taliban tribesmen who are driven to join the terrorist group en masse as a result of "indiscriminate" drone strikes is false. In fact evidence suggests that some of the Pashtuns of FATA who are most exposed to both the drones and the Taliban are the Pakistanis that are most tolerant of the former and intolerant of the latter. A journalist writing about the issue for Reuters reported, "What I have noticed is that at least some among the Pashtun intelligentsia say the drone strikes are precise, and that opposition to them increases the further away you get from the tribal areas."[94] Salman Masood, writing for the *New York Times*, said, "A lot of the people have pointed [out] that there have been lesser protests in the tribal areas over drones as compared to Pakistan proper. The issue has become a bit of a political football."[95] Farhat Taj similarly noticed the tendency for Pakistanis who live across the Indus River in Punjab and Sindh Provinces to be more intolerant of the drone strikes than the Taliban-dominated tribesmen from the FATA region. In an article in the Pakistani newspaper *Dawn*, she wrote,

> There is a deep abyss between the perceptions of the people of Waziristan, the most drone-hit area and the wider Pakistani society on the other side of the River Indus. For the latter, the US drone attacks on Waziristan are a violation of Pakistani's sovereignty. Politicians, religious leaders, media analysts and anchorpersons express sensational clamour over the supposed 'civilian casualties' in the drone attacks. I have been discussing the issue of drone attacks with hundreds of people of Waziristan. They see the US drone attacks as their liberators from the clutches of the terrorists into which, they say, their state has wilfully thrown them. . . .

The people of Waziristan have been complaining why the drones are only restricted to targeting the Arabs. They want the drones to attack the TTP leadership, the Uzbek/Tajik/Turkmen, Punjabi and Pakhtun Taliban. I have heard even religious people of Waziristan cursing the jihad and welcoming even Indian or Israeli support to help them get rid of the TTP and foreign militants. The TTP and foreign militants have made them hostages and occupied their houses by force. The Taliban have publicly killed even the religious scholars in Waziristan.

What we read and hear in the print and electronic media of Pakistan about drone attacks as a violation of Pakistan's sovereignty or resulting in killing innocent civilians is not true so far as the people of Waziristan are concerned. According to them, al Qaeda and the TTP are dead scared of drone attacks and their leadership spends sleepless nights. This is a cause of pleasure for the tormented people of Waziristan.[96]

In an article for the *Daily Times*, Taj came out strongly in support of the drones:

They [the people of Waziristan] want al Qaeda along with the Taliban burnt to ashes on the soil of Waziristan through relentless drone attacks. The drone attacks, they believe, are the one and only "cure" for these anti-civilisation creatures and the US must robustly administer them the "cure" until their existence is annihilated from the world. The people of Waziristan, including tribal leaders, women and religious people, asked me to convey in categorical terms to the US the following in my column.

This was the view of the people of Waziristan. I would now draw the attention of the US to the Peshawar Declaration, a joint statement of political parties, civil society organisations, businessmen, doctors, lawyers, teachers, students, labourers and intellectuals, following a conference on December 12–13, 2009, in Peshawar. The declaration notes that if the people of the war-affected areas are satisfied with any counter-militancy strategy; it is drone attacks that they support the most. Some people in Waziristan compare drones with the Quran's Ababeels—the holy sparrows sent by God to avenge Abraham, the intended conqueror of the Khana Kaaba. . . .

The overpowered people of Waziristan are angry. They believe no one in their entire history has inflicted so much insult on them as al Qaeda. In our native land, they say, al Qaeda has killed so many of us.[97]

Taj made a similar comment in an article posted on Viewpoint:

It is in this context that contrary to the wider public opinion in Pakistan, people of FATA welcome drone attacks and want the Americans to continue hitting the FATA based militants with the drones till their complete elimination. I know all this because of my close association with the area. The same is true about Amn Tehreek, those who passed the Peshawar Declaration as well as the Pakhtun journalists working with radio Mashal. I would encourage the researchers and journalists around the world, who care for professional standards of their work, to also get to know the FATA people's support for the drone strikes through their investigative skills and direct access to people from the drone hit areas.[98]

Even those tribesmen who have given the Taliban sanctuary in their *hujras* out of fear or a feeling of *melmastiia* seem to be tiring of the Taliban. The Taliban's brutal spy witch hunts and subsequent executions, suicide-bombing attacks on civilians, burning of local schools, propensity for violence, policy of forbidding polio vaccinations, and tendency to act as a magnet for drone strikes and Pakistani army invasions has turned many tribesmen against the organization. One tribal elder said of the Taliban, "They are swarming our place. We gave them shelter because we thought they were fighting infidels but now they are dictating what to do in our own land. They set up check posts on the main roads and then ask us about our identity. Who are they to ask us such questions?"[99]

In response, across the FATA tribesmen have formed *lashkars* to fight against the Taliban.[100] The Taliban have responded with suicide bombings of civilian *jirgas* that have killed hundreds in the region, but the battle goes on between the majority who do not support the terrorists and the Taliban.[101] In this sentiment the Pashtuns reflect a trend in their country. In 2008 only 33 percent of Pakistanis held a negative view of the Taliban; by 2009 it had gone up to 70 percent. As for the drones, one Pashtun source known only as "Khan" (to hide his identity from the Taliban) has written, "Another excitement is the sighting of a drone. People and children do not rush indoors, they look at them and discuss and argue about the distance at which they must be flying. The general impression is that they are close. They feel the happiness of something close, friendly and powerful against evil."[102]

As for the notion that Pakistan proper, on the other side of the Indus (Punjab and Sindh), is seething with fury and mass protests against the drones, although it is true that these regions' inhabitants are more inclined to dislike the American campaign than those in the FATA, the antidrone protests have been small and limited. Even Imran Khan, a former world-class cricketeer-turned-politician who has attacked the "hypocrite" Zardari government for allowing the drones, has failed to mobilize the masses based on their hatred for the strikes. A recent article in *Dawn*, titled "US Drone Strikes Fail to Mobilise Pakistan Masses," said, "Campaigners condemn US drone strikes in Pakistan as extra-judicial assassinations that kill hundreds of civilians, but popular protests against them are conspicuous by their rarity. . . . Rallies protesting the CIA-run operation against Taliban and al Qaeda allies in Pakistan's tribal areas on the Afghan border are few and thinly attended." One Pakistani quoted in the article dismissed the efforts of Imran Khan (known as Taliban Khan by his critics) to rally Pakistanis based on violations of sovereignty: "Imran Khan and others are demonstrating against drones and their victims. . . . But can any of these people go to North Waziristan and come back alive?"[103]

The same paradigm can be found in Yemen. As for the idea that drones drive Yemeni tribesmen to join the terrorists, Christopher Swift, an adjunct professor at Georgetown University who carried out fieldwork in Yemen the summer of 2012, found that "to my astonishment, none of the individuals I interviewed drew a causal relationship between drone strikes and Al Qaeda recruiting. Indeed, of the 40 men in this cohort, only five believed that U.S. drone strikes were helping al Qaeda more than they were hurting it." Swift also reported that the primary factors driving young men to join the insurgency in

Yemen were "overwhelmingly economic," not drone-based. A tribal militia commander from one of the provinces that had been taken over by militants who were trying to create a Taliban-style shariah law state summed up his feelings on drones as follows: "Ordinary people have become very practical about drones. If the United States focuses on the leaders and civilians aren't killed, then drone strikes will hurt al Qaeda more than they help them."[104] It would thus seem that in Yemen, as in Pakistan, many who live in the drone-targeted areas have come to have pragmatic views of the drones. It is primarily among Western drone activists and elites in towns in Yemen and Pakistan that the "drones create more terrorists than they kill" paradigm prevails.

Finally, Professor Amitai Etzioni of Georgetown University makes an interesting argument against the "drones make more enemies than they kill" paradigm:

> Such arguments do not take into account the fact that anti-American sentiment in these areas ran high before drone strikes took place and remained so during periods in which strikes were significantly scaled back. Moreover, other developments—such as the release of an anti-Muslim movie trailer by an Egyptian Copt from California or the publication of incendiary cartoons by a Danish newspaper—led to much larger demonstrations. Hence stopping drone strikes—if they are otherwise justified, and especially given that they are a very effective and low-cost way to neutralize terrorist violence on the ground—merely for public relations purposes seems imprudent.[105]

10

The Argument against Drones

*We need to be extremely careful about undermining the
longer-term objective—a stable Pakistan, where elected politicians
control their own national-security establishment, and extremism
is diminishing—for the sake of collecting scalps.*
—**Peter Godspeed, *National Post***

*The problem with the Americans is that the only instrument
up their sleeve is the hammer, and they see everything as a nail.*
—**Anonymous American official quoted in the *Guardian***

During her October 2009 visit to Pakistan, former secretary of state Hillary Clinton was frequently criticized in conferences by Pakistanis who strongly resented the killing of their compatriots on Pakistani soil by Americans prosecuting the war on terrorism. On one occasion an angry Pakistani audience member told Clinton that the drone strikes amounted to a form of "execution without trial."[1] The *Pakistani Observer* said, "Instead of tactical gains or strategic advantage, the daily slaughter of some militants, heavy collateral damage of civilian lives, homes and property will leave long lasting scars, which will never heal."[2] The majority of Pakistanis seem to agree that the distrusted Americans are carrying out a campaign of extrajudicial execution of their countrymen in a unilateral hunt for anti-American terrorists.

DRONE STRIKES ARE A PUBLIC RELATIONS AND STRATEGIC DISASTER IN PAKISTAN

The prevalent Pakistani belief that the majority of those who are being executed by drones are civilians only deepens the distrust of America. A 2010 Pew opinion poll in Pakistan found that "there is little support for U.S. drone strikes against extremist leaders—those who are aware of these attacks generally say they are not necessary, and overwhelmingly they believe the strikes kill too many civilians." Specifically the Pew report stated, "Nearly all (93%) of those who are familiar with the strikes say they are a bad thing. Most Pakistanis (56%) who have heard about the drone attacks say they are not necessary to defend Pakistan from extremist groups, while about one-in-three (32%) believe they are necessary. Nine-in-ten think these attacks kill too many innocent people."[3] A subsequent 2011 Pew poll found that the number of Pakistanis who viewed the drone strikes negatively had risen to 97 percent.[4] A 2012 Pew poll found that "about 75 percent of Pakistanis surveyed regard the United States as an enemy. . . . A key reason for the ongoing ill will appears to be America's use of drone strikes as a tactic against Islamist militants based in Pakistan."[5]

Unlike the poll by the Aryana Institute (discussed in chapter 9), which demonstrated support for the drone strikes among tribesmen in the FATA, a survey by the New America Foundation found the opposite. The New America Foundation reported, "More than three-quarters of FATA residents oppose American drone strikes. Indeed, only 16 percent think these strikes accurately target militants; 48 percent think they largely kill civilians and another 33 percent feel they kill both civilians and militants."[6] Although members of the Aryana Institute have argued that Taliban intimidated many average tribesmen into speaking out against the drone strikes when polled by outsiders, it is also clear that some people in Pakistan proper and the FATA strongly oppose the drone strikes. Their main concern is that the strikes kill too many civilians.

This was a concern I noticed while conducting research in Pakistan in 2010. Although many Pakistanis supported the killing of terrorists—just so long as it was done cleanly—they felt that there was no such thing as an "acceptable" number of civilians being killed in the process. For this reason, most thought the drone strikes were bad for Pakistan. Although, as has been pointed out in previous chapters, the drone campaign is unprecedentedly accurate and leads to relatively few civilian deaths, this was not the perception in Pakistan. Perception can be more important than reality. I found that even anti-Taliban, English-speaking secular elites in Islamabad, Peshawar, and Lahore believed that the drones were killing more civilians than terrorists. They could not tolerate the idea of a distrusted foreign intelligence service killing large numbers of Pakistani men, women, or children who were uninvolved with terrorism, even by accident as collateral damage.

With the Pakistani media banging a steady drumbeat of anti-Americanism, Americans have little power to change this perception. The three separate U.S. studies, discussed in chapter 8, that demonstrate that the drones kill only a small percentage of civilians in their strikes have not altered the Pakistani perceptions that the CIA is brutally killing large numbers of civilians in their country. Few Pakistanis are aware of these studies, and even if they were, they would probably distrust them because they were conducted by Americans. When U.S. officials such as Hillary Clinton visit Pakistan to engage the Pakistani people and present America's softer side, they are drowned out by the voices asking about civilian deaths in drone strikes. This makes it impossible to "sell" America to the Pakistani people.

America is clearly losing the war of perceptions and with it the war for the hearts and minds of millions of Pakistanis, and the drones strikes don't help. The false number of "700 dead civilians for just 14 terrorists" propagated by the antidrone voices in Pakistan is typical of this exaggerated rhetoric. This disinformation is the public relations collateral damage of the drone war, and it may far outweigh the tactical gains that clearly come from the killing of hundreds of Taliban and al Qaeda operatives and the disruption of their terror-insurgency campaign. In its most benign form this growing distrust of the United States and its drone campaign simply leads to anti-American rallies and American flag burnings. At its worse it can lead to Pakistanis, both Pashtuns and non-Pashtuns in places like the Punjab, joining or actively supporting the militants. The drones that kill terrorists may thus be inadvertently recruiting new ones to replenish their ranks. Few issues excite the fury of the Pakistanis more than stories of innocent Pakistani children killed in their homes by drone strikes; this can incite issues of *ghairat* (honor) and *badal* (revenge).

In the larger sense this failure in the war of perceptions undermines not just the Americans' image but also the image of the Pakistani government, which is tied to it. Most Pakistanis see the Zardari government as either complicit in the murder of fellow citizens or too weak to prevent the bullying Americans from carrying out the drone assassination campaign. The Zardari government is forced to continually release public statements criticizing the drone strikes as violations of sovereignty in order to come off as defenders of Pakistan's territorial integrity. The revelation that the CIA drones were being secretly flown from the Pakistani air base at Shamsi in southeastern Pakistan, with the obvious compliance of Pakistani authorities, seriously undermined the government's credibility with its own people. Many Pakistanis felt that the government, which had issued many public criticisms of the drone strikes in the past, was being duplicitous.

For this reason, although there are clearly prodrone voices in the Pakistani government, as seen in chapter 8, one cannot write off all the official protests against the drone campaign as mere pro forma sop for the Pakistani masses designed to put daylight between Islamabad and the infamous CIA drones. The Pakistanis, for example, complained about the strikes to the U.S. ambassador to Pakistan and the head of CENTCOM, Gen. David Petraeus, during a 2008 visit. After a meeting with Petraeus, Pakistani president Zardari said, "Continuing drone attacks on our territory, which result in loss of precious lives and property, are counterproductive and difficult to explain by a democratically elected government. It is creating a credibility gap."[7] This statement is hard to contest. The drone strikes make the weak Pakistani president look bad before his people.

The Pakistani defense minister claimed the strikes were generating "anti-American sentiments" and creating "outrage and uproar among the people." Another military official said that the missile strikes were "counterproductive" and "driving a wedge between the government and the tribal people."[8] Certainly there have been mass protests against the drone strikes, especially following the Damadola strikes. These may have turned local tribesmen against the government and certainly caused an uproar throughout the country.

Pakistani prime minister Yousuf Gilani described the strikes as "disastrous" and said, "Such actions are proving counter-productive to efforts to isolate the extremists and militants from the tribal population."[9] He also said, "We are trying to separate militants from tribesmen, but the drone attacks are doing exactly the opposite."[10] On another occasion he stressed, "The political and the military leadership have been very successful in isolating the militants from the local tribes. But once there is a drone attack in their home region, they get united again. This is a dangerous trend, and it is my concern and the concern of the army. It is also counterproductive in the sense that it is creating a lot of anti-American sentiment all over the country."[11]

In other words, this high-ranking Pakistani official felt the strikes might align aggrieved tribes that lost civilians as collateral damage in drone strikes with the Taliban, which would be catastrophic for the wars in both the FATA and Afghanistan. The Pakistani army's Maj. Gen. Athar Abbas similarly claimed that the missile strikes "hurt the campaign rather than help."[12] And Abdul Basit, a Pakistan foreign office spokesman, expressed his opposition to the strikes saying, "As we have been saying all along, we believe such attacks are counter-productive. They involve collateral damage and they are not helpful in our efforts to win hearts and minds."[13]

This statement is an understatement when it comes to the damage done to America's image in Pakistan. One has only to extrapolate how Americans would feel about the CIA killing real or suspected American extremists without trials in the United States (much less a foreign Muslim intelligence agency such as the ISI doing the same thing) to see how most Pakistanis feel about America's secretive drone assassination campaign in their country. On several occasions Pakistanis anecdotally told me they "liked to know what was going on in their own backyard," and this phrase seemed to have begun to circulate in regard to the drones. The secretive nature of the drone strikes and the CIA's lack of accountability to anyone in its own government (much less the Pakistani government) disturbed many Pakistanis I met.

Thus, powerful figures in Pakistan see the strikes as undermining the country's fragile civilian government and creating problems with the tribesmen who are caught between the drones and the Taliban. The undermining of the already unstable Pakistani government has grave strategic implications. For example, it is difficult for the Pakistani military to carry out its own anti-Taliban operations in places such as North Waziristan because the drone campaign makes it appear as if the army is doing so only in furtherance of U.S.

goals. Pakistanis perceive their army as a "stooge" fighting fellow Muslims on behalf of the Americans.

The drone strikes also provide the Taliban with more reason to hate the Pakistani military and the United States. According to one Pakistani army officer quoted in the London *Times*, the drone strikes have provided the Taliban with a "huge motivation to fight against the Government and the army." Another Pakistani general said, "We complained about it [the strike]. It was detrimental to our operations. I was about to mount an operation and the moment the drone did its attack I had to change dates. Our success lies with the writ of the Government and our popularity with the people. We have to take into account the influences and perceptions these people have."[14] Major General Abbas complained, "The US might have achieved tactical gains through the drone strikes, but they too had caused enormous damage to Pakistan's efforts towards fighting the terrorism."[15] Similarly, progovernment tribal elders have pleaded for an end to the strikes because they "made them look like puppets" and "gave lie to the argument that we've made for a long time, this fight is theirs too."[16] Michael J. Boyle, an assistant professor of political science at La Salle University, summed up the previous Pakistani positions aptly: "Despite the fact that drone strikes are often employed against local enemies of the governments in Pakistan and Yemen, they serve as powerful signals of these governments' helplessness and subservience to the United States and undermine the claim that these governments can be credible competitors for the loyalties of the population."[17]

Criticism of the drone strikes is not limited to the Pakistani government. There has been no person more critical of the strategic setbacks caused by the drone campaign than retired admiral Dennis C. Blair, a former director of national intelligence. Since being forced out of his post in 2010 for voicing his criticism of the strikes, Blair has called for the CIA to hand over the drone campaign to the military. He went so far as to suggest that the United States "pull back on unilateral actions . . . , except in extraordinary circumstances." He further said, "We're alienating the countries concerned, because we're treating countries just as places where we go [to] attack groups that threaten us. We are threatening the prospects for long-term reform raised by the Arab Spring . . . which would make these countries capable and willing allies who could in fact knock that threat down to a nuisance level."[18] Speaking specifically about the drones, Blair said,

> As the drone campaign wears on, hatred of America is increasing in Pakistan. American officials may praise the precision of the drone attacks. But in Pakistan, news media accounts of heavy civilian casualties are widely believed. Our reliance on high-tech strikes that pose no risk to our soldiers is bitterly resented in a country that cannot duplicate such feats of warfare without cost to its own troops.
>
> Our dogged persistence with the drone campaign is eroding our influence and damaging our ability to work with Pakistan to achieve other important security objectives like eliminating Taliban sanctuaries, encouraging Indian-Pakistani dialogue, and making Pakistan's nuclear arsenal more secure.[19]

Blair has also said that there is "little point in killing easily replaceable foot soldiers if the cost is public outrage in Pakistan." He believes that the reliance on unpopular drone strikes undermines America's credibility in Pakistan and hurts the Pakistani army's ability to gain support to fight the war to seize territory from the Taliban. The only way to keep from further alienating the Pakistanis, in Blair's opinion, is to "put two hands on the trigger," that is, allow the Pakistanis veto rights and a voice in choosing drone targets. The war against the Taliban, Blair thinks, cannot be won from the air; it can best be won with aid and assistance to impoverished villages in the FATA. This would help improve the U.S. and Pakistani government's images and win over tribesmen who might be on the fence by offering positive instead of negative incentive. Blair also complained, "The steady refrain in the White House that 'This is the only game in town'—reminded me of body counts in Vietnam."[20]

Blair was not alone in his views. One critic of the drone campaign, Nathaniel Fick of the Center for a New American Security, wrote, "Drone strikes excite visceral opposition across a broad spectrum of Pakistani opinion. The persistence of these attacks on Pakistani territory offends people's deepest sensibilities, alienates them from their government, and contributes to Pakistan's instability."[21] Writing for the *Daily Times*, former Pakistani general Talat Masood similarly argued,

For Pakistan, these strikes are a huge embarrassment. An ally is challenging its sovereignty and independence repeatedly and humiliatingly. . . .

A government that is already under criticism and has credibility issues is being made to look helpless in the face of US attacks. The leadership, especially President Asif Zardari, is losing popularity and no one is prepared to take seriously the official condemnations that follow every incident. . . .

This war has to be won through the people's support, and the advantage that a democratically elected government has over a dictatorship is obliterated if the former is seen as helpless against US strikes. In fact, drone strikes are diverting attention from combating insurgency, and anti-Americanism is on the rise. And even if the militants seem to be losing tactically in the short-term, there will be a long-term rise in the number of militants as well as the number of alienated people. There is further negative blowback as the militants hold the government complicit in these attacks.[22]

The U.S. ambassador to Pakistan and other members of the State Department who are tasked with cleaning up the public relations mess after strikes like the Datta Khel attack, which killed scores of villagers, would most likely agree with Mr. Masood. They think that the CIA ignores the huge diplomatic cost that comes from strikes that now increasingly kill mere Taliban foot soldiers.[23] Opinion in Pakistan, a country of 190 million people, is being turned against the United States all for the sake of killing hundreds of low-level Taliban fighters. The public opinion fallout has given anti-American politicians from the various Islamic political parties a platform to mobilize people against the pro-American Zardari government. Whereas Imran Khan was unable to gather large numbers to his antidrone rallies, Pakistan's main Islamist party, the Jamiat e Ulema, was able to bring together 100,000 people for an antidrone rally in Karachi in January 2012.[24] This sort of mass protest seriously undermines the Zardari government, which is already struggling with the military, the judiciary, other less pro-American political parties, and of course the Taliban and other extremists. Should the weak Zardari government be removed from power by one of these many antidrone groups, the U.S.-Pakistani alliance in the war on terrorism could end.

Many people who oppose the drone strikes also consider them a Whac-a-Mole-type short-term solution that cannot solve the problem of Taliban control in North Waziristan and elsewhere. Wars cannot be won from the air; they have to be fought on the ground. This means that sooner or later the United States will have to rely on Pakistan to ultimately solve the problem of the terrorist sanctuary in the FATA. One Pakistani has warned, "Drone attacks are 'not an effective long-term strategy.' This is an ideological and political war that cannot be won through the use of drones. Each time it is proclaimed that a top militant has been killed, another militant comes up to take up the leadership. Look how after Baituallah's death, Hakimullah took over the reins of the Pakistani Taliban and the militants are as deadly as ever."[25]

DRONES ARE NOT PERFECT; THEY CAN (AND DO) MAKE MISTAKES THAT LEAD TO CIVILIAN DEATHS

Although, as demonstrated in chapter 8, drones are incredibly precise, they are far from perfect killing machines. Drones have killed the wrong people. Examples of such mistakes that can be proven provide ammunition for those who claim that the United States is reliant on techint and humint that is all too fallible.

The first example of a drone operator killing an innocent civilian was in the attack against Mir Ahmad, discussed in chapter 4. Ahmad was a tall Afghan who collected scrap metal in the hills of Zawhar Kili, Afghanistan, during the early days of Operation Enduring Freedom. He was spotted by a drone operator who assumed that anyone that tall in Afghanistan had to be bin Laden. Thus, Ahmad and his friends were blown to bits in an instant with a Hellfire missile. The drone pilots essentially arrogated for themselves the right to be judge, jury, and executioner, and in the process they killed several innocent villagers, entirely as the result of their supreme reliance on technology. No local sources on the ground verified that the target the drone operators had randomly stumbled across on a typical hill in the Texas-sized country of Afghanistan was bin Laden.

Similarly misguided attacks, in which drone pilots have spotted and killed someone whose pattern-of-life movements mistakenly gave him the signature of a Taliban militant, have likely occurred in the FATA. Although the majority of drone kills are supported by both solid, on-the-ground humint and technical intelligence, mix-ups are bound to happen owing to the distances involved and inevitable communication problems. In these cases, innocent people die.

This point was vividly demonstrated with the infamous spring 2011 Datta Khel strike, which, as discussed in chapter 8, took place the day after CIA contractor Raymond Davis was released from a Pakistani jail where he had been detained after he killed two Pakistanis. In this strike as many as fifteen respected tribal elders were killed in a single attack against a local Taliban commander. Although the commander and several of his guards were also killed, the collateral damage among civilians was larger and had far greater ramifications than the killing of the few Taliban militants. An eyewitness account of the strike provides harrowing insight into what it is like for civilians to be attacked by drones:

> The assembly, a traditional Pathan jirga [tribal council], was being held in the open, on flat ground close to the Tochi river, on the Pakistani side of the Afghan border in tribal North Waziristan. There were more than 150 present, gathered to resolve a dispute over how much revenue each of several neighboring clans was due from a chromite mine on the slopes of a nearby mountain.
>
> Sharbat Khan, the contractor who had leased the mining rights, had just begun to speak when four or five Predators—American pilotless "drone" aircraft—flew over the line of brown, craggy hills at the valley's rim and seemingly filled the sky.
>
> Their first target was a car, which was heading away from the Afghan border, being driven along the rough mountain road at high speed in an effort to outrun the drones and their deadly payload. According to witnesses, the aircraft fired four missiles at the car, but it was going so fast that they missed. Then, as the vehicle passed the village of Datta Khel, where the jirga had assembled, the drones fired two more missiles. This time, the car turned into a fireball, and all five men inside were killed.

It may well be that whoever was piloting the drones thousands of miles away, sitting at a computer screen somewhere in America, did have reliable intelligence that the men in the car were terrorists. It is probable, say Pakistani security sources, that a GPS chip had been secreted inside the vehicle by an agent working for the Americans in order to track it more accurately.

But after the car's destruction, and before the tribesmen could take cover, the drones came back and started firing indiscriminately at them. "Four missiles were fired on the jirga members, who included people from all ages," a tribesman, Samiullah Khan, told a local Pathan journalist. "The next moment there was nothing except the bodies of the slain and injured all around." According to Samiullah Khan, the victims' families had to be satisfied with burying disconnected "pieces of flesh." In all, 41 died immediately, and a further seven over the following week.[26]

One local described the aftermath of the strike: "There were pieces—body pieces—lying around. There was lots of flesh and blood." The mourning people of the village were forced to "collect pieces of flesh and blood and put them in a coffin."[27] Unsurprisingly, the reaction among villagers who had lost their respected elders in the notorious strike ranged from sorrow to vows of *badal*-style revenge. One surviving tribesman said, "Our whole village was orphaned because all the elders were killed." A second villager warned, "It has been a big mistake to target the *jirga* as it will have severe consequences."[28] Another similarly stated, "It will create resentment among the locals and everyone might turn into suicide bombers."[29] Finally, a surviving elder said, "Americans don't spare us—not our children, nor our elders, nor our younger. That is why we have decided we will take blood revenge however we can." The remaining elders wrote a statement titled "Announcement of Jihad against America": "We have given permission to our loved ones to do suicide attacks against Americans. And we will take revenge so that Americans will remember it for centuries."[30]

Similar anecdotal evidence suggests errant strikes that kill civilians, at the worst, drive surviving tribesmen into the arms of the militants or, at best, undermine progovernment tribal leaders. Aamir Latif writes,

Until last month, Habibullah was one of many Pakistani tribesmen who considered the Taliban and their foreign operatives as the prime reason for their woes.

But three days after President Barack Obama took the oath of office, everything changed for the 26-year-old. A missile that Habibullah believes was fired by a U.S. drone hit his house, killing his two brothers and a mentally retarded relative…

Now, Habibullah has become a Taliban militant himself, swearing to avenge the deaths of his brothers in line with a centuries-old Pashtun custom of *badal,* or revenge.

Many Pakistani tribesmen resent the Taliban for the self-declared Islamic rules it has imposed on the local population, as well as its backing for foreign operatives living in the tribal regions. But the increasing number of U.S. drone attacks, coupled with bombing raids by Pakistani forces, have made it harder for many to oppose the Taliban's presence. . . . According to Hazar, whenever the tribal elders, and local religious leaders, who have been sidelined by the Taliban, manage to create an anti-Taliban environment, a U.S. drone attack or bombing by Pakistani jets often ruins their efforts.[31]

Striking a similar note, one Pashtun tribesman said, "Many people who did not support the Taliban previously support them now because the Americans are killing innocent people."[32]

Just as several Pakistani government officials suspected, drone strike mistakes had the ability to drive tribesmen to anti-American militancy. In a earlier statement that could have been scripted to fit the scenarios previously described Prime Minister Gilani warned,

The political and the military leadership have been very successful in isolating the militants from the local tribes. But once there is a drone attack in their home region, they get united again. This is a dangerous trend and it is my concern and the concern of the army. It is also counterproductive in the sense that it is creating a lot of anti-American sentiment all over the country. But in order to fight the militants in Waziristan we have to carry the public with us. One cannot go into any war without the support of the masses. We need huge public support to combat terrorism. But we do not get that if there is American interference which we do not ask for.[33]

Gilani's fears that the drone strikes might drive the tribes into the arms of the terrorists were not overblown. On several occasions the Taliban or enraged Pasthun tribesmen have retaliated for air strikes, most notably the Chenagai strike, with suicide bombings. One account of this sort of trend reads, "My neighbor was so furious when a drone killed his mother, two sisters and his 7-year-old brother last September that he filled his car with explosives and rammed it into a Pakistani army convoy. He had to avenge the death of his loved ones."[34] A Pashtun from the region similarly argued that the killing of innocent people was driving anti-Americanism to new highs. According to Mohammad Kamran Khan, "I recently visited North Waziristan during Eid. People were angry with me for the large number of civilians killed in these attacks. They were angry with the Pakistan government and our armed forces for not doing anything to put a halt to these attacks. Also, their hatred towards America was at an all-time high."[35]

A member of the Pakistani parliament said, "The lava of anger and hatred is flowing in the tribal areas. Clearly, the drone attack strategy is not winning people over. It is only increasing hatred against the US and now more people are taking up arms."[36]

The same trend of increased support for terrorists following collateral damage death from drone attacks has been observed in Yemen. One Yemeni businessman said, "The attacks are making people say 'we believe Al Qaeda is on the right side.'" Another Yemeni whose nephew was killed as a bystander in a drone strike stated, "The Americans are targeting the sons of the Awlak. I would fight even the devil to exact revenge for my nephew."[37] A third Yemeni said, "Dear Obama, when a US drone missile kills a child in Yemen, the father will go to war guaranteed. Nothing to do with Al Qaeda."[38] After a drone strike in Yemen wounded seven civilians, an outraged Yemeni said, "Our lives are valueless in the eyes of our government, and that is why civilians are being killed without a crime."[39]

An anguished survivor of another drone strike, which killed a group of Yemeni civilians, including a mother, her seven-year-old daughter, and a twelve-year-old boy, stated, "Their bodies were burning. How could this happen? None of us were Al Qaeda."[40] Another survivor of the strike said, "I would try to take my revenge. I would even hijack an army pickup, drive it back to my village and hold the soldiers in it hostages. I would fight along al-Qaeda's side against whoever was behind this attack." Yet another said, "Our entire village is angry at the government and the Americans. . . . If the Americans are responsible, I would have no choice but to sympathize with al-Qaeda because al-Qaeda is fighting America." And finally, a Yemeni who lost a family member in the strike added, "If there's no compensation from the government, we will accept the compensation from al-Qaeda. If I am sure the Americans are the ones who killed my brother, I will join al-Qaeda and fight against America." Summing up such emotions, a Yemeni source said, "Every time the American attacks increase, they increase the rage of the Yemeni people, especially in al-Qaeda-controlled areas. The drones are killing al-Qaeda leaders, but they are also turning them into heroes."[41]

The fallibility of the CIA's humint and techint and its potentially fatal results are best demonstrated by the case of two U.S. servicemen, Navy Corpsman Benjamin Rast and Marine Staff Sgt. Jeremy Smith, who were mistakenly killed by a drone in a friendly fire incident in Afghanistan. Rast and Smith were misidentified as "Taliban" by a drone flown from Nevada and killed in a salvo of missiles.[42] If the drone operators could not identify fellow Americans on a battlefield in Afghanistan, then there is a high probability they could similarly mistake other armed men in the remote FATA (a region where arms are prevalent) for Taliban.

A similar incident took place in Oruzgan Province, Afghanistan, when drone pilots noticed three trucks, which they mistook for a Taliban convoy, and attacked them with missiles. Unfortunately, the trucks were packed with civilians, and as many as twenty-three noncombatants were incinerated in an instant by drone Hellfires. A U.S. general who investigated the attack on the civilians found that "information that the convoy was anything other than an attacking force was ignored or downplayed" by the overly eager Predator crew whose reporting was "inaccurate and unprofessional."[43] In fact, the drone operators followed the civilian convoy of trucks for three and half hours analyzing its pattern-of-life movement before firing on it and massacring its members. It was only after the strike that drone pilots noticed terrified women survivors waving clothing to surrender to their airborne attackers as they carried babies from the convoy's wreckage.[44] It is likely that in both of these instances of misidentification trigger-happy drone operators gave into the urge to use their technology to mistakenly kill fellow Americans and unarmed Afghan civilians without full humint support. It can be theorized that the same thing has happened in Pakistan as well.

Several other drone errors have cost innocent people their lives. There is, for example, the case of Jabr al Shabwani, a deputy provincial governor in Yemen who met with a local al Qaeda leader to arrange a truce in October 2010. During the meeting a drone fired its missiles into the gathering, killing the popular Shabwani, five of his bodyguards, and the al Qaeda leader. According to Reuters, "The killing so angered Shabwani's tribesmen that in the subsequent weeks they fought heavily with government security forces, twice attacking a major oil pipeline in Maarib."[45] Once again a drone strike proved to be a recruiter for anti-American militancy.

Another strike that demonstrates that the drones are only as good as their ground intelligence took place in Turkey. On November 22, 2011, the United States flew its last Predator and Reaper drones out of Iraq but transferred four of them to Turkey (smaller surveillance drones were, however, kept in Iraq to protect the massive U.S. embassy complex in Baghdad). At the time the Turkish government announced that it would be in charge of the four Predator drones' operations in their country. It was assumed that the drones, which were based at Incirlik Air Base in southeastern Turkey, would be used to monitor the Iraqi-Turkish border. Specifically, they would monitor the infiltration of Kurdish guerrillas coming into Turkey from Iraq to fight for independence against Turkish troops.

On December 29, 2011, one of the drones spotted what appeared to be a group of Kurdish insurgents sneaking across the border from Iraq to Turkey.[46] It then transmitted their location to Turkish F-16 fighter jets, which bombed the group approximately fifteen minutes later. At least thirty-five were killed in the strike. But it later became known that far from being Kurdish guerrillas, the men who were attacked were simply Kurdish smugglers sneaking into Turkey with cigarettes and fuel. Restless Kurds throughout Turkey staged mass protests over the killing of the smugglers, and the case once again proved that for all the technology at their fingertips, the drone pilots were not infallible. It was the largest Kurdish civilian death toll in a single strike in Turkey's three-decade-long war with the Kurdish insurgents.

Sky News reported the aftermath of the errant strike as follows: "Television images showed a line of corpses covered by blankets on a barren hillside, with a crowd of people gathered around—some with their heads in their hands and crying. People loaded the bodies onto donkeys which were led down the hill to be loaded into vehicles and taken to hospital in the mainly Kurdish southeast of the country. Security sources said the people killed had been carrying canisters of diesel on mules and that their bodies were found on the Iraqi side of the border." A local mayor said, "We have 30 corpses, all of them are burned. The state knew that these people were smuggling in the region. This kind of incident is unacceptable. They were hit from the air."[47] To compound matters, most of those who were killed in the strike were Kurdish teenagers whose fathers belonged to a clan that actually fought for the Turkish government against the Kurdish insurgents.[48] This fact suggests that the decision to carry out the deadly strike on the Kurdish smugglers was made without the benefit of any supporting ground humint whatsoever.

A similar errant strike took place in Radaa, Yemen, in September 2012 and led to the death of thirteen civilians. In this case, the drones hit a civilian vehicle traveling near a targeted terrorist vehicle. At the time the Yemeni government apologized for the mistake and said, "This was one of the very few times when our target was completely missed. It was a mistake, but we hope it will not hurt our anti-terror efforts in the region." Grieving family members tried unsuccessfully to carry the bodies of the slain victims to the capital. When the government blocked their passage, a Yemeni who was near the strike angrily said, "You want us to stay quiet while our wives and brothers are being killed for no reason. This attack is the real terrorism."[49]

In her groundbreaking 2009 article on the cost of the drone war for the *New Yorker*, Jane Mayer wrote, "Cut off from the realities of the bombings in Pakistan, Americans have been insulated from the human toll, as well as from the political and the moral consequences. Nearly all the victims have remained faceless, and the damage caused by the bombings has remained unseen."[50] Several well-known civilian victims, however, have provided us faces and names to represent all the innocents who have died in the drone attacks. Chief among them was a young antidrone activist named Tariq Khan. Following is a BBC account of his death:

> When tribal elders from the remote Pakistani region of North Waziristan travelled to Islamabad last week to protest against CIA drone strikes, a teenager called Tariq Khan was among them. A BBC team caught him on camera, sitting near the front of a tribal assembly, or jirga, listening carefully.
>
> Four days later the 16-year-old was dead—killed by one of the drones he was protesting against. In his final days, Tariq was living in fear, according to Neil Williams from the British legal charity, Reprieve, who met him at the Jirga. "He was really petrified," said Mr Williams, "and so were his friends. He didn't want to go home because of the drones. They were all scared."
>
> Tariq carried with him the identity card of his teenage cousin, Asmar Ullah, who was killed by a drone. On Monday he shared his fate. Tariq's family says he was hit by two missiles as he was driving near Miranshah, the main town in North Waziristan. The shy teenager, who was good with computers, was decapitated in the strike. His 12-year-old cousin Wahid was killed alongside him.
>
> The boys were on their way to see a relative, according to Tariq's uncle, Noor Kalam, who we reached by phone. He denied that Tariq had any link to militant groups. "We condemn this very strongly," he said. "He was just a normal boy who loved football."[51]

In the aftermath of the deaths of Tariq Khan and his twelve-year-old cousin, an unnamed U.S. official said of the strike that killed him, "On that day no child was killed; in fact, the adult males were supporting al-Qaeda's facilitation network and their vehicle was following a pattern of activity used by al-Qaeda facilitators."[52]

Another civilian casualty of the drone campaign was Saadullah, whose death was reported as follows:

> Many senior commanders from the Taliban and al-Qaeda are among the dead. But campaigners claim there have been hundreds of civilian

victims, whose stories are seldom told. A shy teenage boy called Saadullah is one of them. He survived a drone strike that killed three of his relatives, but he lost both legs, one eye and his hope for the future. "I wanted to be a doctor," he told me, "but I can't walk to school anymore. When I see others going, I wish I could join them."

Like Tariq, Saadullah travelled to Islamabad for last week's jirga. Seated alongside him was Haji Zardullah, a white-bearded man who said he lost four nephews in a separate attack. "None of these were harmful people," he said. "Two were still in school and one was in college." Asghar Khan, a tribal elder in a cream turban, said three of his relatives paid with their lives for visiting a sick neighbor. "My brother, my nephew and another relative were killed by a drone in 2008," he said. "They were sitting with this sick man when the attack took place. There were no Taliban."[53]

Another similar civilian death was featured in Salon.com in 2010:

Gul Nawaz, from North Waziristan, was watering his fields when he heard the explosion of drone missiles: "I rushed to my house when I heard the blast. When I arrived I saw my house and my brother's house completely destroyed and all at home were dead." Eleven members of Gul Nawaz's family were killed, including his wife, two sons and two daughters as well as his elder brother, his wife, and his four children. "Yes, the drone strikes hurt the Taliban. Most of the strikes are effective against the Taliban but sometimes innocent people also become the victim of such attacks. Take my case," said Gul Nawaz.

"I blame the government of Pakistan and the USA, they are responsible for destroying my family. We were living a happy life and I didn't have any links with the Taliban. My family members were innocent. . . . I wonder, why was I victimized.[54]

Other civilian victims of drone strikes have been covered in the media. For example, Noor Behram, a resident of North Waziristan, has spent years photographing the civilian aftermath of drone strikes in his native region. He has organized an exhibit of his photographs in London and a similar one, titled "Bugsplat," was held in Lahore. Following are some of the captions that accompanied photographs Behram published:

Dande Darpa Khel (North Waziristan), Aug. 21, 2009

The stench that Behram smelled when he arrived at Dande Darpa Khel came from the charred bodies of Bismullah Khan and his wife. Near the bombed-out remains of their house, Behram found the Khans' three living children. The children—the younger two girls on the left, their older brother on the right—were in shock, and clutched the ruins of their neighbor's house as if the rubble could comfort them. "These kids had no idea where their parents were. They didn't know their parents were killed," Behram says. Also killed in the blast: their brother, Syed Wali Shah, age 7. Behram later heard that the children were taken in by their uncle. "There's no government here, no social network or security," he explains. "People have to look after each other."

Dande Darpa Khel, Aug. 21, 2009

By the time Behram reached Bismullah Khan's mud house, partially destroyed in the strike, Khan's youngest son, Syed Wali Shah, had already died. Behram watched as the boy's body was laid out on a prayer rug, a "very small" one, in preparation for his funeral. "The body was whole," Behram recalls. "He was found dead." The villagers wrapped a bandage around the boy's head, even though they had no chance to save his life. Behram doesn't know who the target of the Dande Darpa Khel attack was. ("You'd have to ask the CIA that," he says.) But he observed people's anger as they prepared bodies for burial and cleared the wreckage. "The people were extremely angry. They were talking and shouting against the U.S. for the attack," Behram says.

Datta Khel, Oct. 18, 2010

Pakistan's *Express Tribune* reported a drone attacked "two suspected militant hideouts" in Datta Khel near Mirin Shah. Behram never saw the scene. He headed instead to a Mirin Shah hospital, where he heard residents had frantically driven one of the strike's victims: Naeemullah, a boy of about 10 or 11. Naeemullah was said to be injured in the strike after a missile struck the house next door. Shrapnel and debris travelled into Naeemullah's house, wounding him in his "various parts of his body," Behram says. "You can't see his back, but his back was wounded by missile pieces and burns." An hour after Behram took this picture, Naeemullah died of his injuries.[55]

While Behram admits that he did not take pictures of dead Taliban victims (who were likely removed from the scenes by fellow militants), he does not necessarily have to present the whole picture. The pictures and stories of dead Pakistani civilians are enough to confirm to Pakistanis that the U.S. drones do in fact kill civilians. As for the locals' responses to the drone attacks, the *Guardian* writes,

> According to Noor Behram, the strikes not only kill the innocent but injure untold numbers and radicalize the population. "There are just pieces of flesh lying around after a strike. You can't find bodies. So the locals pick up the flesh and curse America. They say that America is killing us inside our own country, inside our own homes, and only because we are Muslims.
>
> "The youth in the area surrounding a strike gets crazed. Hatred builds up inside those who have seen a drone attack. The Americans think it is working, but the damage they're doing is far greater."[56]

Another account of a drone strike reads, "Their bodies, carbonized, were fully burned. They could only be identified by their legs and hands. One body was still on fire when he reached there. Then he learned that the charred and mutilated corpses were relatives of his who lived in his village, two men and a boy aged seven or eight. They couldn't pick up the charred parts in one piece."[57] Regarding the acceptability of civilian casualties, another Pakistani angrily said, "I think, even if they said, 'we've killed 100 terrorists,' and just one child was also killed. . . . If you, at that time, you see that child's body, you talk to his mother and father—I think, for me, this is a very serious thing. That one child, sitting in his house, could be killed like this."[58]

Researchers at Stanford University and New York University have compiled a report of stories from drone victims that brings to life those whose lives have been shattered by drone strikes. One account reads,

"Before the drone strikes started, my life was very good. I used to go to school and I used to be quite busy with that, but after the drone strikes, I stopped going to school now. I was happy because I thought I would become a doctor." Sadaullah recalled, "Two missiles [were] fired at our *hujra* and three people died. My cousin and I were injured. We didn't hear the missile at all and then it was there."

He further explained, "[The last thing I remembered was that] we had just broken our fast where we had eaten and just prayed. . . . We were having tea and just eating a bit and then there were missiles. . . . When I gained consciousness, there was a bandage on my eye. I didn't know what had happened to my eye and I could only see from one."

Sadaullah lost both of his legs and one of his eyes in the attack. He informed us, "Before [the strike], my life was normal and very good because I could go anywhere and do anything. But now I am not able to do that because I have to stay inside. . . . Sometimes I have really bad headaches . . . [and] if I walk too much [on my prosthetic legs], my legs hurt a lot. [Drones have] drastically affected life [in our area]."[59]

There are also cases of people who allowed armed Taliban, often uninvited, into their houses and then paid the price for the visit with a drone strike. For example, the Pakistani paper the *News* reported,

The Taliban and Al Qaeda have unleashed a reign of terror on the people of FATA. People are afraid that the Taliban will suspect their loyalty and behead them. Thus, in order to prove their loyalty to the Taliban and Al-Qaeda, they offer them to rent their houses and hujras for residential purposes.

There are people who are linked with the Taliban. Terrorists visit their houses as guests and live in the houses and hujras. The drone attacks kill women and small children of the hosts. These are innocent deaths because the women and children have no role in the men's links with terrorists.

Other innocent victims are local people who just happen to be at the wrong place at the wrong time.[60]

A similar case was reported in the *Asia Times*: "In an interview with a researcher for CIVIC, a civilian victim of a drone strike in North Waziristan carried out during the Obama administration recounted how his home had been visited by Taliban fighters asking for lunch. He said he had agreed out of fear of refusing them. The very next day, he recalled, the house was destroyed by a missile from a drone, killing his only son."[61]

Although it is common knowledge in the FATA that the drones are hunting militants, not civilians, many Pashtuns fear that they will be killed by accident, and so they live in a state of fear. A Pakistani journalist interviewed one tribesman named Khaista Khan: "The people in North Waziristan, currently the main target of the drone strikes, are developing psychological disorders because of the constant fear and anxiety caused by the drones regularly flying over the area. 'Everyone is scared here,' Khan said. 'It is like someone is pointing a loaded gun at you when you are working, eating your meal, sitting with the children or sleeping. It is becoming very difficult to live this way.'"[62]

A journalist for the *Miami Herald* painted a similar picture of the tribesmen's fear of drones:

> They described a terrifying existence under the drones in North Waziristan, the focus of the strikes. A 13-year-old boy, Saddam Hussain, said that he lost his 10-month-old niece and sister-in-law in a strike on their house on the night of Oct. 9, in the Datta Khel area of North Waziristan. Saddam carried a large picture of the baby girl with him at the protest. "The drones patrol day and night. The sound comes when they fly lower down. Sometimes we see six in the air all at once," Saddam said. "When they come down, people run out of their houses, even at night."[63]

A *Globe and Mail* reporter in the FATA reported,

> People who sleep under the buzzing of the drones say it's hard to settle down for the night, listening to the sound of armed machines nearby. Muhammad Amad, executive director of Idea, an aid group that works in the tribal areas, was telling a visitor that the drones are counterproductive because they stir up local anger, when he was inter-

rupted by one of his local staffers from Waziristan, interjecting in broken English: "Mental torture," said the bearded man, with sun-weathered skin. He repeated himself, struggling to enunciate: "Mental torture." "Yes, it's mental torture," Mr. Amad said. "When we lie down under the noise of the drones, nobody sleeps."

Several people from the tribal areas said the same thing. Sleeping pills and anti-depressants have become a regular part of the diet, they said, even in poor villages where few people can afford meat.[64]

The Xinhua news agency similarly reported, "Dr. Faizur Rehman Burki, a local physician, said that the drone strikes have not only panicked people, but also catalyzed uncertainty because of which people were now using sedatives. 'Usage of tranquilizers has been increased,' Xinhua news agency quoted Dr. Burki, as saying. 'I am not scared, but haunted by the uncertainty that anything can happen anytime to my home and the loved ones,' said Naseemullah, a native of Wana."[65]

A local source told the *Los Angeles Times*, "These drones fly day and night, and we don't know where to hide because we don't know who they will target. If I could, I would take revenge on America."[66] Another Pakistani said, "People are very worried, very tense all the time. When the missile is fired from the plane, there is a loud explosion. When it hits the ground, it makes a terrifying noise. The people below, they just start running. Pieces of missile, they fly everywhere, very far, into other people's houses."[67]

Similar sentiments have been expressed in Yemen, where CIA drones have made several strikes on al Qaeda operatives, including the previously mentioned attack that accidentally killed a popular Yemeni governor. According to one Yemeni source, "The drones fly over Marib every 24 hours and there is not a day that passes that we don't see them. The atmosphere has become weary because of the presence of U.S. drones and the fear that they could strike at any time."[68]

Thus it would seem that for every voice in favor of the drones found in the previous chapter, there is a voice against them. Some supporters in the targeted regions support the drone campaign whereas others live in fear of the unseen killers in the sky. Some root for the drones to kill the militants who terrorize them whereas others point out that the drones are self-defeating in that they act as accidental recruiters for these very same terrorists when they kill civilian bystanders. That this is the case should not be surprising considering the controversial nature of any assassination campaign carried out during a time of war—especially one being run by a distrusted foreign government's covert intelligence agencies.

11

The Future of Killer Drones

The development of a new generation of military robots,
including armed drones, may eventually mark one
of the biggest revolutions in warfare in generations.
—**Anna Mulrine, *Christian Science Monitor***

It's a good time to be a flying robot.
—**Spencer Ackerman and Noah Shachtman, *Wired***

T here can be no doubt that drones represent the future of counter-terrorism and counterinsurgency in remote, unpoliced lands, such as Pakistan's FATA region, Yemen, Somalia, and Libya. Where U.S. troops cannot be placed on the ground, drones will increasingly fly to strike at those whom America deems to be its enemies. In 2011, during a speech given at Harvard University, John Brennan, the president's top counterterrorism adviser, announced, "The United States does not view our authority to use military force against al-Qaeda as being restricted solely to 'hot' battlefields like Afghanistan."[1] This means that the U.S. government believes it can use drones wherever al Qaeda may be, from the Maghreb in North Africa (where al Qaeda in the Islamic Maghreb operates) to Mindanao in the southern Philippines (home to several pro–al Qaeda Islamic groups, such as Abu Sayyaf). All signs are that the U.S. military and the CIA are planning a future in which drones play an increasingly important role in warfare and antiterrorist operations.

This of course means more strikes in Afghanistan and Pakistan, the primary focus of current drone operations. As the United States draws down its troops in Afghanistan in 2014 and prepares to hand the fight against the Taliban over to allied Afghan National Army and Afghan National Police troops, its ground presence in this strategic country will be much diminished, and the Pentagon will turn over much of the reduced American combat against the Taliban insurgents to small, elite groups of rapid-reaction special operations troops, manned support aircraft, and of course, drones. These troops and drones, which will most likely be based in residual bases, or so-called Joint Facilities, in Jalalabad (eastern Afghanistan), Kandahar (southern Afghanistan), and Bagram (north of Kabul), will be used to assist the Afghan army in repelling Taliban swarm assaults on town centers and will bolster the Afghan army's efforts to carry out offensives against Taliban-held sanctuaries. They will also engage in "hunt and kill" missions designed to take out local Taliban commanders and disrupt their networks.

Most importantly, the Pakistani Taliban and Afghan Taliban have reportedly made an agreement to unite their forces to fight not against the Pakistani government but to overthrow the pro-U.S. government in Afghanistan.[2] The Taliban alliance will be emboldened by the drawdown of U.S. forces in Afghanistan after 2014 and will doubtless widen their operations inside Afghanistan. As the Taliban and its al Qaeda allies try to carve out sanctuaries in Afghanistan, drone strikes will increasingly be necessary to keep them from openly gathering and exerting authority à la the FATA model.

The withdrawal of the majority of U.S. troops in Afghanistan will also bolster support for al Qaeda in the FATA. The need for counterterrorism-counterinsurgency personality strikes in this de facto Taliban statelet will be greater than ever. Pakistani major general Shafiq Ahmed has presciently stated, "If America wants to stay in Afghanistan, or safeguard its interests in case of a proposed pull-out [from Afghanistan], it has to tame North Waziristan."[3] This will certainly mean a continuation of signature-strike attacks on Taliban foot soldiers as well.

The drones will also play a key role in keeping up the pressure on AQAP in Yemen and al Shabab in Somalia. The new Yemeni president, Abdu Rabu Mansour Hadi, has condoned the strikes against the terrorists and insurgents who took advantage of the turmoil following the 2011 downfall of the Saleh government to carve out sanctuaries in the remote Abyan Province. In 2012 there were forty-two strikes in Yemen, almost as many as in Pakistan that year (forty-six).⁴ And in Somalia, U.S. special operations troops and drones are increasingly being used both to raid Shabab militants and to monitor pirates who have seized Western captives.

Libya provides an example of future uses for drones. As mentioned previously, in 2011 there were more drone strikes in Libya during the overthrow of Gaddafi than there were in Pakistan. The *Global Post* described this Libyan campaign as the model for future drone campaigns: "The death [of Gaddafi] is the latest victory for a new American approach to war: few if any troops on the ground and the heavy use of air power, including drones."⁵ Regarding the drawbacks of conventional warfare, as opposed to drone campaigns, Micah Zenko, a fellow at the Council on Foreign Relations, writes, "The lessons of the big wars are obvious. The cost in blood and treasure is immense, and the outcome is unforeseeable. Public support at home is declining toward rock bottom. And the people you've come to liberate come to resent your presence."⁶

The drone-centric alternative to "big wars" dovetails with the Pentagon's and CIA's long-term plans for counterterrorism and counterinsurgency operations in the Islamic world and beyond. Former CIA official Bruce Riedel has said of Obama's plans, "This administration has made a very conscious decision that it wants to get out of large conventional warfare solutions and wants to emphasize counterterrorism and a lighter footprint on the ground."⁷ Obama has said that the U.S. military of the future will focus on "intelligence, surveillance, reconnaissance, counterterrorism, countering weapons of mass destruction, and the ability to operate in environments where adversaries try to deny us access."⁸ All these tasks can be done by drones. As Vice President Joe Biden put it in the October 2012 vice presidential debate, "We don't need more M1 tanks, what we need is more UAVs."⁹

Although the recent economic crunch has led to huge cuts in the U.S. military's budget (the Pentagon is making $487 billion in cuts over ten years, eliminating at least eight brigades, and reducing the size of the active army from 570,000 to 490,000 troops), the Pentagon is set to increase its drone combat air patrols from sixty-one to eighty-five. It has called for a 30 percent increase in the drone fleet in coming years.¹⁰ This represents a shift from big bloody wars, like the invasion of Iraq, which cost more than $1 trillion and forty-five hundred American lives, to the aerial campaign in Libya, which cost just more than $1 billion and no U.S. lives.

In addition to bases in Turkey, Sicily, Afghanistan, and (potentially once again) Pakistan, drones will be launched from forward staging bases that some advisers are calling "lily pad bases." These bases include those currently found in Camp Lemonier, Djibouti, and Arba Minch, Ethiopia.[11] Similar new bases may be built in Jordan and Turkey to help monitor Iraq and in the Seychelles Islands of the eastern coast of Africa to hunt Somali pirates.[12] President Obama also authorized the building of a new secret drone base in the Rub al Khali Desert in eastern Saudi Arabia to carry out strikes on AQAP.[13] In response to the takeover of northern Mali by extremists from Ansar Dine, al Qaeda in the Islamic Maghreb, and allied Tuareg rebels in the winter of 2012–2013, the president called for the creation of a drone base in Niamey, the capital of neighboring Niger.[14] Obama's most recent defense budget calls for funding for the construction of an "afloat forward staging base," that is, a launching pad for drones and special operations units that can be sailed around the world to potential hot spots.[15] This base could park offshore and send CIA or JSOC drones into nearby countries to kill targets without having to ask the local government for permission.

The U.S. intelligence community keeps its plans more secret than the military, but these drone basing trends certainly reflect the CIA's drone future as well. In fact, their drones have already flown from lily pad bases in Ethiopia, Djibouti, Pakistan, Saudi Arabia, and perhaps elsewhere. The CIA division that controls the drones, the Counterterrorism Center, has grown from three hundred employees to two thousand since 9/11 and now represents about 10 percent of the agency's workforce.[16] Thus the CIA, which once focused more on espionage, will doubtless continue to carry out is counterterrorism drone operations in all the previously described contexts.

THE DRONE REVOLUTION

Whether one supports the drone strikes or is opposed to them, there is no doubt that drones are here to stay. A few facts about drones will make their permanence abundantly clear:

- In 2000 the United States had just fifty drones. Today almost one in three U.S. warplanes is a drone. That translates to approximately 7,500 drones in the U.S. fleet. The majority of them (5,346) are

Ravens, a small hand-launched surveillance drone used by the army.[17] Nearly every brigade that fought in Iraq or Afghanistan had a Raven for "look down" or "overwatch" surveillance purposes.

- In October 2012 the CIA asked the White House for ten more drones to add to its already existing fleet of as many as thirty-five. There has been discussion of deploying these additional drones in North Africa against al Qaeda in the Islamic Maghreb, against militants in post-Gaddafi Libya, and in the vast expanses of northern Mali that were briefly conquered by al Qaeda–linked militants in the winter of 2012–2013.[18]

- Since 2005 patrols by drones have increased 1,200 percent.[19]

- The Air Force trained more drone pilots in 2011 than regular pilots.[20] More than half of all undergraduate pilot training graduates are assigned to pilot drones rather than manned aircraft.[21] Since 2008 the number of Air Force drone pilots has grown fourfold to eighteen hundred.[22]

- In nine years the Pentagon has increased its drone fleet thirteen-fold and is spending $5 billion a year adding to it.[23]

- A recent Defense Department plan calls for a 30 percent increase in the size of the U.S. drone fleet in coming years.[24]

- In August 2011 the United States revealed it would be investing around $23 billion in advancing its drone program. This at a time of steep military cutbacks.[25]

- Since 2001 the military has spent more than $26 billion on drones.[26]

- Globally over the next decade more than $94 billion is expected to be spent on drone research and procurement.[27]

- British military officials have said that almost one-third of Royal Air Force aircraft will be drones in twenty years.[28]

- More than fifty countries have built or bought drones. Even the Lebanese terrorist group Hezbollah has used Iranian-built drones. Many observers are worried about a future drone arms race that will see countries other than the United States hunting down their enemies with remote-control planes.[29]

- The UK has developed a $225 million jet-propelled drone capable of hitting targets on other continents. Known as the Taranis, it was named after the Celtic god of thunder.[30] Unlike the ungainly Predator and Reaper, the stealth technology–equipped Taranis has an internal bomb bay that can carry a wide array of weapons.

- In March 2013 General Atomic Aeronautical Systems agreed to sell $197 million worth of drones to the United Arab Emirates, in the first sale of drones to a non-NATO member. The unarmed version of the Predator is to be known as the Predator XP and will be used for surveillance missions.[31]

- The Pentagon intends to spend approximately $37 billion on a variety of drones including the MQ-9 Reaper and the Global Hawk, a high-flying drone spy plane.[32]

- The U.S. military currently has sixty-five advanced MQ-9 Reapers, and it plans to receive four hundred more of them.[33]

- The U.S. military has procured more than 250 MQ-1 Predator drones.[34]

- The 2011 defense budget sought funds for a 75 percent increase in drone operations.[35]

- The military plans to buy more than eighty Global Hawk surveillance drones, which cost $141 million per aircraft.[36]

- The U.S. Navy is developing a carrier-based jet drone known as the X-47B, which can fly ten times farther than manned planes and defend aircraft carriers from threats such as "carrier killer" missiles.[37]

- The United States recently launched a surveillance drone, known as the Phantom Eye, that can remain aloft for four days gathering intelligence.[38]

- The U.S. Air Force is developing nanodrones, such as the Wasp, that weigh less than a pound and can fly to a thousand feet. The Air Force has also planned Project Anubis (named for the Egyptian god of death) to build small killer drones that weigh less than a pound. The small drones will be used to terminate HVTs and could one day fly in swarms against an enemy.[39]

- The U.S. Army recently developed a small backpack-size drone, known as the Switchblade. This kamikaze aircraft carries explosives that can be launched from a tube, loiters in the sky, and dives at targets upon command.[40]

- China unveiled twenty-five new drone models at an air show in 2011, and Iran claims to have two drones, known as "messengers of death," that are capable of long-range missions.[41]
- By late 2011 U.S. drones had logged 2.7 million hours of flight with the majority of that time (87 percent) being flown in combat.[42]
- The U.S. Army has developed a surveillance drone that can be flown by the crew of an Apache AH-64D Longbow attack helicopter to help it find its targets on the ground.[43]
- Predator drones are already being used to monitor the U.S.-Mexican border. Recently a Mexican police drone crashed in the United States.[44]
- America has already experienced its first attempt by a terrorist to use a drone to carry out a terrorist act. In September 2011 Rezwan Ferdaus was arrested in the Boston area after the Federal Bureau of Investigation (FBI) found him plotting to use seven-foot remote-control toy planes loaded with C-4 plastic explosives to blow up the Pentagon and other targets in Washington, DC.[45]
- Palestinian sources say more than eight hundred people have been killed by Israeli drone strikes in the Gaza strip in recent years.[46]
- The U.S. Air Force has begun purchasing a new jet drone known as the Predator C, or Avenger, that will allow it to deliver munitions to a target at a much faster speed than the propeller-driven Predators and Reapers in its current fleet. The Avenger carries even more ammunition than the Reaper.[47]
- Since the United States began its drone war in Pakistan, more than two thousand people have been killed by U.S. drones.[48] That is more than the total U.S. combat loses in Afghanistan in a decade of fighting.
- In December 2010 the U.S. Air Force announced that it had test-flown the X-37B, a drone modeled on the space shuttle, into space. This development caused many drone critics to worry that the Air Force was involved in weaponizing drones for space warfare.[49]
- In February 2012 the NATO alliance agreed to deploy a fleet of its own drones after seeing how useful the American drones were in

the joint NATO-U.S. air war on Gaddafi forces in Libya.[50] NATO has already begun building a 1.3 billion Euro drone base at Sigonella in Sicily.[51]

Weaponized and surveillance drones are clearly the future of American counterterrorism and counterinsurgency operations, and perhaps even of conventional warfare. Whereas the first drone attack on al Qaeda, which took place in Yemen in 2002, was greeted with tremendous coverage by the international media, drone strikes today are considered so mundane that they are now relegated to small articles on newspapers' back pages, if they are picked up at all. The vast majority of Americans, both Democrats and Republicans, seems to have accepted this radical development with little real debate. In fact 83 percent of Americans approve of Obama's stepped up drone policy, including liberal Democrats, 77 percent of whom support the president on the issue of drone strikes.[52] For Americans, drone attacks in distant locations seem to be an accepted part of the post-9/11 world.

Despite the CIA's reluctance to enter the drone assassination business prior to 9/11, former CIA head David Petraeus once said, "We can't get enough drones."[53] In 2010 former defense secretary Robert Gates said, "We are buying as many Reapers as we possibly can."[54] That same year the commander of the 147th Reconnaissance Wing, Col. Ken Wisian, said of drones, "The demand for this kind of capacity is insatiable."[55]

With little discussion, the United States (along with as many as fifty other nations) has inaugurated what amounts to a drone revolution. Although the CIA is the only intelligence agency in the world that currently flies killer drones beyond its borders to hunt terrorists and insurgents, it is perhaps only a matter of time before Russia, China, Israel, and other countries deploy drone fleets abroad in search of their foes.[56] David Cortright of Notre Dame has fretted, "What kind of a future are we creating for our children? We face the prospect of a world in which every nation will have drone warfare capability, in which terror can rain down from the sky at any moment without warning."[57]

As Cortright and others ponder the future of remote-control aerial killers and their impact on war and counterterrorism, drones are increasingly coming to shape the way the United States and other countries hunt and kill those they deem to be enemies. Peter Singer, author of *Wired for War: The Robotics Revolution and Conflict in the Twenty-First Century*, best sums up the new drone reality: "The [drone] technology is here. And it isn't going away. It will increasingly play a role in our lives. . . . The real question is: How do we deal with it?"[58]

APPENDIX: DRONE SPECIFICATIONS

MQ-1 PREDATOR

Primary Function: Armed reconnaissance, airborne surveillance, and target acquisition

Contractor: General Atomics Aeronautical Systems Inc.

Power Plant: Rotax 914F four-cylinder engine

Thrust: 115 horsepower

Wingspan: 55 feet (16.8 meters)

Length: 27 feet (8.22 meters)

Height: 6.9 feet (2.1 meters)

Weight: 1,130 pounds (512 kilograms) empty

Fuel Capacity: 665 pounds (100 gallons)

Speed: Cruise speed around 84 miles per hour (70 knots), up to 135 miles per hour

Range: Up to 770 miles (675 nautical miles)

Ceiling: Up to 25,000 feet (7,620 meters)

Armament: Two laser-guided AGM-114 Hellfire missiles

Crew: Two (pilot and sensor operator)

Unit Cost: $20 million (fiscal 2009 dollars; includes four aircraft, a ground control station, and a Predator primary satellite link)[1]

MQ-9 REAPER

Primary Function: Remotely piloted hunter-killer weapon system

Contractor: General Atomics Aeronautical Systems Inc.

Power Plant: Honeywell TPE331-10GD turboprop engine

Thrust: 900 shaft horsepower maximum

Wingspan: 66 feet (20.1 meters)

Length: 36 feet (11 meters)

Height: 12.5 feet (3.8 meters)

Weight: 4,900 pounds (2,223 kilograms) empty

Maximum takeoff weight: 10,500 pounds (4,760 kilograms)

Fuel Capacity: 4,000 pounds (602 gallons)

Payload: 3,750 pounds (1,701 kilograms)

Speed: Cruise speed around 230 miles per hour (200 knots)

Range: 1,150 miles (1,000 nautical miles)

Ceiling: Up to 50,000 feet (15,240 meters)

Armament: Combination of AGM-114 Hellfire missiles, GBU-12 Paveway II, and GBU-38 JDAMs

Crew: Two (pilot and sensor operator)

Unit Cost: $53.5 million (fiscal 2006 dollars, includes four aircraft with sensors)

Initial Operating Capability: October 2007[2]

NOTES

1. THE DEATH OF A TERRORIST

1. Declan Walsh, "Is Baitullah Mehsud Now Public Enemy No 1 for the US?" *Guardian*, April 5, 2009.
2. "41 Dead in Pakistan Suicide Bombing: Officials," Agence-France Presse, December 25, 2010, http://www.thejakartaglobe.com/afp/41-dead-in-pakistan-suicide-bombing-officials/413540; "Pakistan Assessment 2012," South Asia Terrorism Portal, http://www.satp.org/satporgtp /countries/pakistan/ (accessed on February 19, 2013); Shahan Mufti, "Suicide Attacks a Growing Threat in Pakistan," *Christian Science Monitor*, October 10, 2008.
3. Sabrina Tavernise, "Deaths at Hands of Militants Rise in Pakistan," *New York Times*, January 14, 2009.
4. "Pro-Taliban Commander Threatens Benazir with Suicide Attacks," *AndhraNews*, October 5, 2007, http://www.andhranews.net/Intl/2007/October/5/Taliban-commander-18068.asp; Bill Roggio, "Pakistan Implicates Baitullah Mehsud in Bhutto Assassination," *Long War Journal*, December 28, 2007, http://www.longwarjournal.org/archives/2007/12/pakistan _implicates.php. See also Philip Reeves, "Did Baitullah Mehsud Kill Benazir Bhutto?" *All Things Considered*, National Public Radio, January 16, 2008, http://www.npr.org/templates /story/story.php?storyId=18159635.
5. "Taliban Commander Baitullah Mehsud," *Newsweek*, April 3, 2009.
6. Bill Roggio, "Taliban Capture over 100 Pakistani Soldiers in South Waziristan," *Long War Journal*, August 31, 2007, http://www.longwarjournal.org/archives/2007/08/taliban _capture_over.php.
7. Imtiaz Ali and Craig Whitlock, "Taliban Commander Emerges as Pakistan's 'Biggest Problem,'" *Washington Post*, January 10, 2008.
8. White House, "Remarks by the President on a New Strategy for Afghanistan and Pakistan," press release, March 27, 2009, http://m.whitehouse.gov/the_press_office/Remarks-by-the-President-on-a-New-Strategy-for-Afghanistan-and-Pakistan/.
9. "Scenic Pakistani Valley Falls to Taliban Militants," Associated Press, December 29, 2008.
10. Salman Masood, "Video of Flogging Rattles Pakistan," *New York Times*, April 4, 2009. For the unbearably gruesome videos of the beheadings see "Video: Taliban Slaughter Pakistani Police and Residents Like Animals," *Islamization Watch* (blog), April 22, 2009, http://islamizationwatch.blogspot.com/2009/04/video-taliban-slaughter-pakistani.html.
11. "High-Profile Victories in the Battle against Terror," *Sunday Times* (London), August 9, 2009.
12. "Profile: Hakimullah Mehsud," BBC News, May 3, 2010, http://news.bbc.co.uk/2/hi /south_asia/8219223.stm.
13. Imtiaz Gul, *The Most Dangerous Place: Pakistan's Lawless Frontier* (New York: Viking, 2010), 15.

14. "Taleban, We Will Launch an Attack in Washington That Will Amaze Everyone in the World," *Sunday Times*, April 1, 2009.

15. Bill Roggio and Alexander Mayer, "Charting the Data for US Airstrikes in Pakistan, 2004–2011," *Long War Journal*, http://www.longwarjournal.org/pakistan-strikes.php (accessed in December 2011).

16. Bill Roggio, "Scores of Taliban Killed in Second US Strike in South Waziristan," *Long War Journal*, June 23, 2009, http://www.longwarjournal.org/archives/2009/06/seventeen_taliban_ki.php.

17. Nick Schifrin, "Near Miss: CIA Drone Almost Hits Taliban Chief in Pakistan," ABC News, June 26, 2009, http://abcnews.go.com/Blotter/International/story?id=7939317.

18. "Taliban Leader Baitullah Mehsud Escapes U.S Missile Strike," *Trends Updates*, June 24, 2009.

19. "US Drone Kills Scores of People in Tribal Zone," *Times* (London), June 25, 2009.

20. "Missile Attack Kills Fifty in South Waziristan," *Dawn*, June 24, 2009.

21. Nick Schifrin, "CIA Drone Strike Kills Wife of Pakistani Taliban Chief," ABC News, August 5, 2009, http://abcnews.go.com/Blotter/story?id=8258637&page=1.

22. Ismail Khan, "Good Riddance, Killer Baitullah," *Dawn*, August 8, 2009.

23. Ibid.

24. "Baitullah Mehsud Dead," Paklinks.com, August 6, 2009, http://www.paklinks.com/gs/pakistan-affairs/339128-baitullah-mehsud-dead.html.

25. "Taleban Commander Baitullah Mehsud Killed in US Missile Strike," *Times*, August 8, 2009.

26. Jane Mayer, "The Predator War: What Are the Risks of the C.I.A.'s Covert Drone Program?" *New Yorker*, October 26, 2009.

27. Ishtiaq Mahsud, "New Pakistani Taliban Leader Vows Revenge," *SFGate*, October 6, 2009, http://www.sfgate.com/news/article/New-Pakistani-Taliban-leader-vows-revenge-3284609.php.

28. "New Taliban Leader Vows to Carry on War," *Sky News*, September 18, 2009, http://news.sky.com/story/725368/new-taliban-leader-vows-to-carry-on-war.

29. "Pakistan Taliban Says Leader Injured in Attack," CNN, January 15, 2010, http://www.cnn.com/2010/WORLD/asiapcf/01/15/pakistan.taliban/index.html.

2. A HISTORY OF THE PASHTUN TRIBAL LANDS OF PAKISTAN

1. Mary Anne Weaver, *Pakistan: Deep Inside the World's Most Frightening State* (New York: Farrar, Straus and Giroux, 2010), 69.

2. "Pakistan Air Force/Pakistan Fiza'ya," http://www.f-16.net/f-16_users_article14.html (accessed on January 27, 2011).

3. "The Pakistani US Relationship," *Defense Journal*, April 1998.

4. "Frenemies," *Newsweek*, May 23, 2011.

5. Peter Bergen, *Holy War Inc.* (New York: Touchstone, 2001), 73.

6. The Pashtuns make up about 40 percent of Afghanistan.

7. "CIA Tried to Have Bin Laden Killed," *New York Times*, September 30, 2001.

8. "Officials: U.S. Missed Chance to Kill Bin Laden," Associated Press, June 23, 2003.

9. Daniel Schorr, "Reviewing the Presidential Ban on Assassinations," National Public Radio, September 17, 2001, http://www.npr.org/news/specials/americatransformed/essays/010917.schorrcommentary.html.

3. ENTER THE PREDATOR

1. Bill Yenne, *Birds of Prey: Predators, Reapers and America's Newest UAVs in Combat* (North Branch, MN: Specialty Books, 2010), 15.

2. Linda D. Kozaryn, "Predators Bound for Bosnia," American Forces Press Service, U.S. Department of Defense, February 8, 1996, http://www.defense.gov/news/newsarticle.aspx?id=40516.

3. Michael R. Thirtle, Robert Johnson, and John Birkler, *The Predator ACTD: A Case Study for Transition Planning to the Formal Acquisition Process* (Washington, DC: Rand, 1997), 24–25.

4. Steve Coll, *Ghost Wars: The Secret History of the CIA, Afghanistan, and Bin Laden, from the Soviet Invasion to September 10, 2001* (New York: Penguin, 2005), 529.

5. Ibid.

6. Norman Polmar, *The Naval Institute Guide to the Ships and Aircraft of the U.S. Fleet* (Annapolis, MD: Naval Institute Press, 2005), 479.

7. Yenne, *Birds of Prey*, 43.

8. Peter Singer, *Wired for War: The Robotics Revolution and Conflict in the Twenty-First Century* (New York: Penguin Books, 2009), 33.

9. National Commission on Terrorist Attacks upon the United States, *The 9/11 Commission Report: Final Report of the National Commission on Terrorist Attacks upon the United States* (Washington, DC: U.S. Government Printing Office, 2004), 189–90.

10. David Ensor, "Drone May Have Spotted Bin Laden in 2000," CNN, March 17, 2004, http://articles.cnn.com/2004-03-17/world/predator.video_1_bin-senior-al-predator-drone -aircraft; George Tenet, *Written Statement for the Record of the Director of Central Intelligence before the National Commission on Terrorist Attacks upon the United States*, March 24, 2004, 15, http://www.9-11commission.gov/hearings/hearing8/tenet_statement.pdf.

11. Klaus Brinkbaumer and John Goetz, "Obama's Shadowy Drone War," *Der Spiegel*, October 12, 2010.

12. Ibid.

13. Walter Boyne, "How the Predator Grew Teeth," *Air Force Magazine* 92, no. 7 (July 2009): 2–17.

14. Yenne, *Birds of Prey*, 45.

15. Ted Bridis, "Officials: U.S. Slow on Bin Laden Drones," Associated Press, June 24, 2003.

16. Barton Gellman, "A Strategy's Cautious Evolution," *Washington Post*, January 20, 2002.

17. "Officials: U.S. Missed Chance to Kill Bin Laden," Associated Press, June 23, 2003.

18. Ibid.

19. Scott Shane, "CIA to Expand Use of Drones in Pakistan," *New York Times*, December 3, 2009.

20. Coll, *Ghost Wars*, 581.

21. Ibid.

22. Ibid.

23. See National Commission on Terrorist Attacks, *9/11 Commission Report*, chap. 6, 210–14, for a discussion of this debate.

24. George Tenet, *At the Center of the Storm: My Years at the CIA* (New York: HarperCollins, 2007), 143.

25. Mayer, "Predator War."

26. Avery Plaw, *Targeting Terrorists: A License to Kill?* (Burlington, VT: Ashgate, 2008), 2, 117, 119–20, 210.

27. "Officials: U.S. Missed Chance."

28. Singer, *Wired for War*, 35.

4. OPERATION ENDURING FREEDOM

1. Richard Clarke, *Against All Enemies: Inside America's War on Terror* (New York: Free Press, 2004), 23.

2. Bob Woodward and Dan Baltz, "Combating Terrorism: 'It Starts Today,'" *Washington Post*, February 1, 2002.

3. "Bring Me the Head of Bin Laden," BBC News, May 4, 2005, http://news.bbc.co.uk/2/hi /americas/4511943.stm.

4. "The Yemen Strike: The War on Terror Goes Global," *Chicago Tribune*, November 24, 2002.

5. Eric Schmitt and Thom Shanker, *Counterstrike: The Untold Story of America's Secret Campaign against Al Qaeda* (New York: Times Books, 2011), 4.

6. "CIA Gets New Powers to Eliminate Bin Laden as UAVs Get Armed," Agence France-Presse, October 21, 2001.

7. Barton Gellman, "CIA Weighs 'Targeted Killing' Missions: Administration Believes Restraints Do Not Bar Singling Out Individual Terrorists," *Washington Post*, October 28, 2001.

8. Ibid.

9. Michael Warren, "Biden Insinuates He Didn't Vote for Afghanistan, Iraq Wars," *Weekly Standard*, October 11, 2012.

10. Morris Davis, "Combatant Immunity and the Death of Anwar al Awlaqi," *Jurist*, October 17, 2011, http://jurist.org/forum/2011/10/morris-davis-anwar-al-awlaqi.php.

11. Hassan Abbas, *Pakistan's Drift into Extremism* (London: M. E. Sharpe, 2009), 217.

12. Ibid., 218.

13. For an overview of this war see Brian Glyn Williams, *Afghanistan Declassified: A Guide to America's Longest War* (Philadelphia: University of Pennsylvania Press, 2012).

14. "CIA Sent Drone to Save Rebel Leader," Associated Press, October 29, 2001.

15. Brian Glyn Williams, *The Last Warlord: The Life and Legend of the Afghan Warrior Who Led US Special Forces in Toppling the Taliban Regime* (Chicago: Chicago Review Press, 2013).

16. Brian Glyn Williams, "Report from the Field: General Dostum and the Mazar i Sharif Campaign: New Light on the Role of Northern Alliance Warlords in Operation Enduring Freedom," *Small Wars and Insurgencies* 21, no. 4 (December 2010): 610–32.

17. Terry Anderson, *Bush's Wars* (Oxford: Oxford University Press, 2011), 87.

18. "Tribal Leader: Bin Laden in Pakistan," United Press International, April 21, 2004.

19. Tenet, *Written Statement for the Record*, 16.

20. Yenne, *Birds of Prey*, 45.

21. Carlota Gall, "War-Weary Commanders Admit Rift with Leaders," *New York Times*, February 22, 2011.

22. Thomas E. Ricks, "Target Approval Delays Irk Air Force Officers," *Washington Post*, November 18, 2001.

23. Yenne, *Birds of Prey*, 45.

24. Stephen Grey, "Death of Bin Laden's Deputy: How the US Killed Al Qaeda Leaders by Remote Control," *Times*, November 18, 2001.

25. Ricks, "Target Approval Delays."

26. "Predator and Prey," *Newsweek*, January 16, 2006. Among others, the military's drones are flown by the 11th, 15th, 17th, 46th, and 64th Reconnaissance Squadrons and the Air National Guard's 163rd Reconnaissance Wing, 119th and 214th Reconnaissance Groups, 19th and 42nd Attack Squadrons, 53rd Wing, 56th Test and Evaluation Squadron, 27th, 33rd, and 58th Special Operations Wings, 551st Special Operations Squadron, and 174th Fighter Wing,

27. *Operation Anaconda: The Battle of Robert's Ridge, Part 4 of 5*, Military Channel documentary, *LiveLeak.com*, posted by "bravo61," October 12, 2008, http://www.liveleak.com/view?i =151_1223869362.

28. Doug Struck, "Casualties of U.S. Miscalculations: Afghan Victims of CIA Missile Strike Described as Peasants, Not Al Qaeda," *Washington Post*, February 11, 2002.

29. Doug Struck, "Men Hit in U.S. Missile Strike Were Scavengers, Relatives Say Afghans Were at Al Qaeda Site for Scrap Metal to Sell," *Washington Post*, February 12, 2002.

30. Department of Defense, "DoD News Briefing: Secretary Rumsfeld and Gen. Myers," news transcript, February 12, 2002, http://www.defense.gov/transcripts/transcript.aspx ?transcriptid=2636.

31. Struck, "Men Hit in U.S. Missile Strike."

5. MANHUNT

1. U.S. Department of Defense, "Secretary Rumsfeld Interview with CNN Live Today," news transcript, March 8, 2002.

2. "Interviews Former Treasury Secretary Paul O'Neill," *60 Minutes*, CBS, January 11, 2004.

3. David Rennie, "US Tried to Kill Warlord with Drone Missile," *Telegraph* (London), May 10, 2002.

4. Ron Moreau et al., "Day of the Vampire," *Newsweek*, June 17, 2002.

5. Bob Graham, "Senator Bob Graham Remarks to the Council on Foreign Relations" (speech, Council on Foreign Relations, Washington, DC, March 26, 2004), http://www.cfr.org /terrorism/senator-bob-graham-remarks-council-foreign-relations/p6905.

6. Tom Brook, "US Shifts Spy Planes to Afghan War," *USA Today*, August 23, 2009.

7. David Sanger, *The Inheritance: The World Obama Confronts and the Challenges of American Power* (New York: Harmony Books, 2009), 138.

8. "Dogfight between MQ-1 Predator Drone and Mig-25 Foxbat," YouTube video, 2:04, from CBSNews.com, posted by "OakRidgejet," November 19, 2006, http://www.youtube.com /watch?v=wWUR3sgKUV8.

9. Philip Smucker, "The Intrigue behind the Drone Strike," *Christian Science Monitor*, November 12, 2002.

10. "Yemen Strike Opens New Chapter in War on Terror," *Time*, November 5, 2002.

11. Smucker, "Intrigue behind the Drone Strike."

12. James Bamford, *The Shadow Factory: The Ultra-Secret NSA from 9/11 to the Eavesdropping on America* (New York: Doubleday, 2009), 135.

13. "They Didn't Know What Hit Them," *Time*, November 18, 2002.

14. "U.S. Citizen Killed by CIA Linked to N.Y. Terror Case," *Washington Post*, November 9, 2002.

15. "Assassination by Remote Control," *Economist*, November 5, 2002.

16. "U.S. Drone Hits Al-Qaida in Yemen; 6 Alleged Operatives Killed," *Los Angeles Times*, November 5, 2002.

17. Hassan Abbas, "A Profile of Tehrik-i-Taliban Pakistan," *Combating Terrorism Center Sentinel* 1, no. 2 (January 28, 2008): 3.

18. Abdul Saboor Khan, "Orokzai Becomes New Haven for the Taliban," *Daily Times* (Pakistan), February 4, 2009.

19. Owen Bennett-Jones, "Tide Turns against the Taliban," BBC News, May 21, 2009, http://news .bbc.co.uk/2/hi/south_asia/8059900.stm.

20. "Qaeda Firmly Rooted in Tribal Fiefdom, Report," Agence France-Presse, August 26, 2011.

21. "Pakistan: 100 Fighters Captured in Battle: Confusion over Whether Al-Zawahiri Being Protected," CNN, May 6, 2004, http://www.cnn.com/2004/WORLD/asiapcf/03/20 /pakistan.alqaeda/index.html; "49 Troops Dead or Missing So Far in Al Qaeda Offensive," Agence France-Presse, March 23, 2004.

22. "Pakistan: 100 Fighters"; "49 Troops Dead or Missing."

23. "Nek Warns of Attacks throughout Pakistan," *Dawn*, June 11, 2004.

24. Ibid.

25. Ismail Khan, "Night Raid Kills Nek, Four Other Militants: Wana Operation," *Dawn*, June 19, 2004, http://archives.dawn.com/2004/06/19/top1.htm.

26. Ibid.

27. "US, Pakistani Forces Clash on Afghan Border," CNN, January 1, 2003, http://www.cnn.com /2003/WORLD/asiapcf/central/01/01/afghan.clashes/index.html.

28. "CIA Drone Kills Al-Qaida Operative," MSNBC, May 14, 2005, http://www.nbcnews.com /id/7847008/ns/us_news-security/t/cia-drone-kills-al-qaida-operative.

29. "Surveillance Operation in Pakistan Located and Killed Al Qaeda Official," *Washington Post*, May 15, 2005; "Exclusive: CIA Aircraft Kills Terrorist: Senior Al Qaeda Operative Struck by Predator Missile," ABC News, May 13, 2005, http://abcnews.go.com/WNT/Investigation /story?id=755961.

30. David Ensor, "Sources: Key Al Qaeda Operative Killed," CNN, May 14, 2005, http://www.cnn .com/2005/WORLD/asiapcf/05/13/alqaeda.killing/index.html.

31. "CIA Drone Kills Al-Qaida Operative," MSNBC, May 14, 2005, http://www.nbcnews.com /id/7847008/ns/us_news-security/t/cia-drone-kills-al-qaida-operative.

32. "Exclusive: CIA Aircraft Kills Terrorist."

33. Aamir Borbra, "Musharraf Okayed Drones, Not Strikes: PM," *Daily Mail News*, October 23, 2010.

34. "An Extrajudicial Execution by the CIA?" Amnesty International, May 18, 2005, http://www.amnesty.org/en/library/asset/AMR51/079/2005/en/bcffa8d8-d4ea-11dd-8a23 -d58a49c0d652/amr510792005en.html.

35. "CIA Drone Kills Al-Qaida Operative."

36. Bill Roggio, "Cross-Border Strike Targets One of the Taliban's 157 Training Camps in Pakistan's Northwest," *Long War Journal*, August 13, 2008, http://www.longwarjournal .org/archives/2008/08/crossborder_strike_t.php.

37. "Evidence Suggests U.S. Missile Used in Strike," NBC News, December 5, 2005, http://www .nbcnews.com/id/10303175/#.USFX4Gckf5Q.

38. Craig Whitlock and Kamran Khan, "Blast in Pakistan Kills Al Qaeda Commander: Figure Reportedly Hit by U.S. Missile Strike," *Washington Post*, December 4, 2005.

39. B. Raman, "Mystery Shrouds Top Terrorist's Death," *Rediff.com*, December 5, 2005, http://www.rediff.com/news/2005/dec/05raman.htm.

40. "Al Qaeda No. 3 Dead, but How?" CNN, December 4, 2005, http://www.cnn.com/2005 /WORLD/asiapcf/12/03/pakistan.rabia/index.html.

41. Yenne, *Birds of Prey*, 103.

42. Haji Mujtaba, "US Missile Parts at Pakistan Al Qaeda Site," Information Clearing House, December 5, 2005, http://www.informationclearinghouse.info/article11216.htm.

43. For the full story, see the important documentary "The Stories of a Warlord and a Journalist," part 5 of *Return of the Taliban*, *Frontline*, PBS, October 3, 2006, http://www.pbs.org/wgbh/pages/frontline/taliban/view/main.html. See also "Evidence Suggests U.S. Missile Used in Strike."

44. "Stories of a Warlord and a Journalist."

45. "US Missile Parts at Pakistan Al Qaeda Site."

46. Raman, "Mystery Shrouds Top Terrorist's Death."

47. "Blast Kills Al Qaeda Commander," BBC News, December 3, 2005, http://news.bbc.co.uk/2 /hi/south_asia/4494428.stm.

48. "AP Exclusive: Close Calls for Al-Qaida's No. 2," Associated Press, November 29, 2010.

49. Christina Lamb, "Airstrike Misses Al-Qaeda Chief: 'Wrong Information' Blamed for Pakistan Deaths," *Sunday Times*, January 15, 2006.

50. Brian Ross, "U.S. Strike Killed Al Qaeda Bomb Maker," ABC News, January 18, 2006. http://abcnews.go.com/WNT/Investigation/story?id=1517986.

51. "Pakistan: At Least 4 Terrorists Killed in U.S. Strike," *USA Today*, January 17, 2006; "AP Exclusive: Close Calls."

52. Elaine Shannon, "Can Bin Laden Be Caught?" *Time*, January 22, 2006.

53. "AP Exclusive: Close Calls"; Michael Hirsh et al., "Predator and Prey," *Newsweek*, January 29, 2006.

54. Imtiaz Ali, "Pakistan Fury as CIA Airstrike on Village Kills 18," *Telegraph*, January 15, 2006.

55. Ross, "US Strike Killed Al Qaeda Bomb Maker"; "Pakistan: At Least 4 Terrorists."

56. "AP Exclusive: Close Calls."

57. Associated Press, "Pakistanis Protest US Airstrike," Fox News, January 16, 2006, http://www .foxnews.com/story/0,2933,181698,00.html.

58. Lamb, "Airstrike Misses Al-Qaeda Chief."

59. Shannon, "Can Bin Laden Be Caught?" claims that thirteen died in the strike; "Pakistan Condemns Deadly Airstrike Targeting Al Qaeda Number 2," Associated Press, January 15, 2006, reports the number as thirty.

60. "Zawahiri Strike Sparks Protests," BBC News, January 14, 2006, http://news.bbc.co.uk /2/hi/africa/4613108.stm.

61. Associated Press, "Pakistanis Protest US Airstrike," Fox News, January 16, 2006, http://www.foxnews.com/story/0,2933,181698,00.html.

62. "Pakistan Condemns Deadly Airstrike."
63. "Terror Chief Calls Bush a 'Butcher,'" CBS, January 31, 2006, http://www.cbsnews.com /2100-224_162-1255755.html.
64. "Pakistan Condemns Deadly Airstrike."
65. William E. Alberts, "Remember Damadola," *CounterPunch*, March 4–5, 2006, http://www .counterpunch.org/alberts03042006.html.
66. Hirsh et al., "Predator and Prey."
67. Shannon, "Can Bin Laden Be Caught?"
68. Ibid.
69. Greg Miller, "War on Terror Loses Ground: Al Qaeda Is Regrouping in Pakistan, an Ally the U.S. Must Work with but Doesn't Trust," *Los Angeles Times*, July 27, 2008.
70. Bruce Rolfson, "Predator Was Involved in Zarqawi Assault," *Air Force Times*, July 17, 2006.
71. Bill Roggio and Daveed Gartenstein-Ross, "Pakistan Surrenders," *Weekly Standard*, October 2, 2006, http://www.daveedgr.com/news/pakistan-surrenders/.
72. Gordon Corera, "Bomb Plot—The Al-Qaeda Connection," BBC News, September 9, 2008, http://news.bbc.co.uk/2/hi/uk_news/7606107.stm.
73. "Pakistan School Raid Sparks Anger," BBC News, October 30, 2006, http://news.bbc.co.uk /2/hi/6099946.stm.
74. Christina Lamb, "US Carried Out Madrasah Bombing," *Sunday Times*, November 26, 2006.
75. Anwarullah Khan, "82 Die as Missiles Rain on Bajaur: Pakistan Owns Up to Strike: Locals Blame US Drones," *Dawn*, October 31, 2006.
76. Ibid.
77. Yousuf Ali, "Most Bajaur Victims Were under 20," *News* (Pakistan), November 5, 2006.
78. Ibid.
79. Christine Fair, *The Militant Challenge in Pakistan* (Washington, DC: National Bureau of Asian Research, 2010), 125.
80. "Pakistan School Raid Sparks Anger."
81. Dan Gillmor, "Suicide Blast Kills 42 Pakistani Soldiers," *Guardian*, November 8, 2006.
82. "Pakistan Says Planes Used in Raid," BBC News, January 19, 2007, http://news.bbc.co.uk/2 /hi/south_asia/6280153.stm.
83. Bill Roggio, "The Pro-Osama Meeting in Bajaur," *Long War Journal*, October 29, 2006, http://www.longwarjournal.org/archives/2006/10/the_proosama_meeting.php.
84. Associated Press, "Missile Strike in Pakistan Kills 4," *Washington Post*, April 27, 2007. http://www.washingtonpost.com/wp-dyn/content/article/2007/04/27/AR2007042701056 .html.
85. Griff Witte, "Blast Kills at Least 20 in Pakistan," *Washington Post*, June 20, 2007.
86. Barbara Sude, *Al-Qaeda Central: An Assessment of the Threat Posed by the Terrorist Group Headquartered on the Afghanistan-Pakistan Border*, Counterterrorism Strategy Initiative Policy Paper (Washington, DC: New America Foundation, February 25, 2010).
87. Schmitt and Shanker, *Counterstrike*, 100.
88. "Musharraf Vows War on Militants," BBC News, July 12, 2007, http://news.bbc.co.uk/2 /hi/6896179.stm.
89. "Scores Killed in Pakistan Attacks," BBC News, July 19, 2007, http://news.bbc.co.uk/2 /hi/south_asia/6905808.stm.
90. Ismail Khan, "Missile Kills 5 in Northwest Pakistan: U.S. Denies Attack," *New York Times*, November 3, 2007.
91. Katherine Hubbard, *Strategic Concerns, Historical Ties Make Pakistan Unwilling to Take On Haqqani Network* (Washington, DC: Center for Strategic and International Studies, February 3, 2010).
92. Mark Mazzetti and Eric Schmitt, "Pakistanis Aided Attack in Kabul, U.S. Officials Say," *New York Times*, August 1, 2008.
93. Elizabeth Rubin, "In the Land of the Taliban," *New York Times*, October 22, 2006.

94. "Mullah Akhtar Usmani Named Successor to Mullah Omar," *Rediff.com*, November 19, 2001, http://www.rediff.com/us/2001/nov/19ny7.htm.

95. Ahmed Rashid, *Descent into Chaos: The US and the Failure of Nation Building in Pakistan, Afghanistan, and Central Asia* (New York: Viking, 2008), 370.

6. THE DRONE WAR BEGINS

1. Greg Miller and Judie Tate, "CIA Focus Shifts to Killing Targets," *Washington Post*, August 30, 2011.

2. Brian Glyn Williams, "Cheney Attack Reveals Taliban Suicide Bomber Targeting Patterns," *Terrorism Monitor* 5, no. 4 (March 1, 2007): 2–5.

3. "A Strike against Al Qaeda," *Economist*, February 1, 2008.

4. "Al Qaeda's New Leadership," *Washington Post*, September 8, 2007.

5. "Top Al Qaeda Commander Killed," BBC News, February 1, 2008, http://news.bbc.co.uk /2/hi/south_asia/7220823.stm; Jon Boone and Jason Burke, "Top Al Qaeda Leader Killed in Afghanistan," *Guardian*, January 31, 2008.

6. "Missile Strike Near Pakistan Border Kills 13, Injures 7," *Gulf News*, February 28, 2008.

7. Ibid.; "Many Die in Pakistani Missile Strike," Al Jazeera, February 28, 2008.

8. "16 Dead in Pakistan Missile Strike," Agence France-Presse, March 16, 2008.

9. Candace Rondeux, "Airstrike Kills 18 in Pakistan," *Washington Post*, March 17, 2008, http://www.washingtonpost.com/wp-dyn/content/article/2008/03/16/AR2008031600840 .html.

10. Sanger, *Inheritance*, 235.

11. Schmitt and Shanker, *Counterstrike*, 119.

12. Mark Hosenball, "With a Quiet Blessing, U.S. Attacks on Al Qaeda Spike," *Newsweek*, March 22, 2008.

13. David Sanger and Eric Schmitt, "Pakistan Shift Could Curtail Drone Strikes," *New York Times*, February 22, 2008.

14. Sanger, *Inheritance*, 235.

15. David Cloud, "CIA Drones Have Broader List of Targets," *Los Angeles Times*, May 5, 2010.

16. Adam Entous, Siobhan Gorman, and Julian Barnes, "US Tightens Drone Rules," *Wall Street Journal*, November 4, 2011.

17. Sanger, *Inheritance*, 236.

18. Marcy Wheeler, "'Pattern of Life' Drone Strikes," *Empty Wheel* (blog), May 6, 2010, http://emptywheel.firedoglake.com/2010/05/06/pattern-of-life-drone-strikes/.

19. Noah Shachtman, "No-Name Terrorists Now CIA Drone Targets," *Danger Room* (blog), *Wired*, May 6, 2010, http://www.wired.com/dangerroom/2010/05/no-name-terrorists-now -cia-drone-targets/.

20. Schmitt and Shanker, *Counterstrike*, 102–3.

21. "Claim Pakistan 'Supports Taliban,'" BBC News, October 26, 2011, http://www.bbc.co.uk /news/world-south-asia-15456858.

22. Matthew Aid, *Intel Wars: The Secret History of the Fight against Terror* (New York: Bloomsbury, 2012), 108–9.

23. Jeremiah Gertler, *US Unmanned Aerial Systems*, CRS Report for Congress (Washington, DC: Congressional Research Service, January 3, 2012), 35.

24. "MQ-9 Reaper/Hunter Killer UAV," *Defense Update*, July 2009, http://defense-update.com /products/p/predatorB.htm.

25. Tom Vanden Brook, "Faster, Deadlier Pilotless Plane Bound for Afghanistan," *USA Today*, August 27, 2007.

26. "Reaper Scores Insurgent Kill in Afghanistan," *Air Force Times*, October 29, 2007.

27. Anwarullah Khan, "12 Killed in Drone Attack on Damadola," *Dawn*, May 14, 2008.

28. Jason Burke, "Al Qaeda Chief Dies in Missile Strike," *Guardian*, June 1, 2008.

29. Zahid Hussain and Michael Evans, "US Airstrike Kills 11 Pakistani Soldiers in 'Cowardly and Unprovoked Attacks,'" *Times*, June 12, 2008.

30. Bill Roggio, "Pakistani Leader Killed in March 2008 Predator Strike," *Long War Journal*, May 19, 2008, http://www.longwarjournal.org/archives/2009/05/pakistani_al_qaeda_l.php.

31. Declan Walsh, "US Bomb Kills Eleven Pakistani Troops," *Guardian*, June 12, 2008.

32. "US Missile Strike Kills One in S Waziristan," *Daily Times*, June 15, 2008.

33. "Al Qaeda Chemical Expert Killed," BBC News, July 28, 2008, http://news.bbc.co.uk/2/hi/south_asia/7529419.stm.

34. Dan Darling, "Al Qaeda's Mad Scientist," *Weekly Standard*, January 19, 2006.

35. "Pakistan Seeks Confirmation of Top Al-Qaeda Death," Agence France-Presse, July 27, 2008, http://afp.google.com/article/ALeqM5iZeuUhq2BG8kHdGutqrP0ujlye8A.

36. "US Kills Al Qaeda Chief in Air-Strike," *Telegraph*, July 19, 2011.

37. Kathy Gannon, "Al Qaeda Said to Lose Key WMD Operative," *San Diego Union-Tribune*, August 9, 2008. http://legacy.utsandiego.com/news/world/20080809-1135-al-qaidasloss.html.

38. "Taliban Commander, 9 Others Killed in Missile Attack," *Geo.TV*, August 13, 2008, http://www.geo.tv/8-13-2008/22577.htm; and Roggio, "Cross-Border Strike."

39. "Pakistan Missile Strike Kills Eight: Officials," Agence France-Presse, August 20, 2008, http://afp.google.com/article/ALeqM5grLj6xRsGSHp7FJtyLelxa6i3omA.

40. "Pakistan's Musharraf Steps Down," BBC News, August 18, 2008, http://news.bbc.co.uk/2/hi/7567451.stm.

41. Government Accountability Office, *U.S. Efforts to Address the Terrorist Threat in Pakistan's Federally Administered Tribal Areas Require a Comprehensive Plan and Continued Oversight* (Washington, DC: May 2008).

42. "Missile Attack Kills 5 Militants in Waziristan," *Daily Times*, August 31, 2008; "Two Canadians Killed in Wana Missile Attack," *Dawn*, August 31, 2008.

43. Huma Yusuf, "Fallout of the Davis Case," *Dawn*, February 21, 2011.

44. "6 Killed, 8 Injured in Missile Attack in N. Waziristan," *Geo.TV*, September 1, 2008.

45. Farhan Bokhari, Sami Yousufzai, and Tucker Reals, "U.S. Special Forces Strike in Pakistan," CBS News, September 3, 2008, http://www.cbsnews.com/stories/2008/09/03/terror/main4409288.shtml.

46. Pir Zubair Shah, Eric Schmitt, and Jane Perlez, "American Forces Attack Militants on Pakistani Soil," *New York Times*, September 3, 2008.

47. Ibid.

48. Julian Barnes and Greg Miller, "Pentagon May Step Up Raids in Pakistan," *Los Angeles Times*, September 5, 2008.

49. Eric Schmitt and Mark Mazzetti, "Bush Said to Give Orders Allowing Raids in Pakistan," *New York Times*, September 10, 2008.

50. Shah et al.,"American Forces Attack Militants."

51. Jonathan Karl, Nick Schifrin, Kirit Radia, and Luis Martinez, "US Conducts First Raid on Terrorists in Pakistan," ABC News, September 3, 2008, http://abcnews.go.com/International/story?id=5718172.

52. "Pakistani Fury over US Assault," BBC News, September 4, 2008, http://news.bbc.co.uk/2/hi/7597529.stm.

53. "Bush OKs US Raids into Pakistan—Gen. Kayani Vows to Protect Sovereignty," Arabnews.com, September 12, 2008.

54. Jane Perlez, "Pakistan's Military Chief Criticizes U.S. over a Raid," *New York Times*, September 10, 2008.

55. AFP, quoted in Bill Roggio, "Report: US Airstrike Kills 4 in North Waziristan," *Long War Journal*, September 4, 2008, http://www.longwarjournal.org/archives/2008/09/report_us_airstrike.php.

56. Pir Zubair Shah and Jane Perlez, "U.S. Missiles Killed at Least Six People on Afghanistan-Pakistan Border, Residents Say," *New York Times*, September 5, 2008.

57. Jane Perlez and Pir Zubair Shah, "US Attack on Taliban Kills 23 in Pakistan," *New York Times*, September 9, 2009.
58. "Guard, Al Qaeda Chief in Pakistan Killed," CNN, September 9, 2008, http://www.cnn.com /2008/WORLD/asiapcf/09/09/pakistan.alqaeda.killed/index.html.
59. "'Another US Strike' Hits Pakistan," BBC News, September 12, 2008, http://news.bbc.co.uk/2 /hi/south_asia/7611721.stm.
60. Ibid.; "Suspected US Missile Kills Twelve in Pakistan," *Standard*, September 12, 2008.
61. Ibid.
62. Anwar Iqbal, "Drone Strikes Kill High Value Targets, US Tells Pakistan," *Dawn*, February 9, 2009.
63. "Seven Killed in Missile Strike in S. Waziristan," *Geo.TV*, September 17, 2008, http://www.geo.tv /9-17-2008/25085.htm; "Pakistan Helps US Set Up Missile Strike: Source," Reuters, September 17, 2008.
64. Jeremy Scahill, "The Secret US War in Pakistan," *Nation*, November 23, 2009.
65. Julian Barnes and Greg Miller, "Pakistan Gets a Say in Drone Attacks on Militants," *Los Angeles Times*, May 13, 2009.
66. Scahill, "Secret US War in Pakistan."
67. Nicholas Schmidle, "Getting Bin Laden, What Happened That Night in Abbottabad," *New Yorker*, August 8, 2011.
68. Ken Dilanian, "US Intensifies Drone Strikes in Pakistan," *Los Angeles Times*, September 28, 2010.
69. Noah Shachtman, "US Military Joins CIA's Drone War in Pakistan," *Danger Room* (blog), *Wired*, December 10, 2009, http://www.wired.com/dangerroom/2009/12/us-military-joins -cias-drone-war-in-pakistan/.
70. "2008: President Zardari Was Willing to Take Heat of 'Cross-Border Raid,'" *Dawn*, May 26, 2011.
71. "US Drone Strike Kills Six People," *Telegraph*, October 1, 2008.
72. "US Missile Kills 20 in N. Waziristan," *Geo.TV*, October 4, 2008, http://www.geo.tv/10-4 -2008/26211.htm.
73. "US Missile Kills 6 in Pakistan," Reuters, October 9, 2008.
74. "US Drone Kills 5 in North Waziristan," *Geo.TV*, October 12, 2009.
75. Pir Zubair Shah, "US Strike Is Said to Kill Al Qaeda Figure in Pakistan," *New York Times*, October 17, 2008.
76. Pir Zubair Shah, "Airstrike Kills 8 Militants in Pakistan, Villagers Say," *New York Times*, October 23, 2008.
77. "Suspected US Strike Kills Eight in South Waziristan," *Geo.TV*, October 27, 2008, http://www .geo.tv/10-27-2008/27631.htm.
78. "'US Strike' Kills Taleban Leader," BBC News, October 27, 2008, http://news.bbc.co.uk/2 /hi/south_asia/7692373.stm.
79. "Suspected Missile Kills Seven in S. Waziristan," *Geo.TV*, October 31, 2008, http://www.geo.tv /10-31-2008/27916.htm.
80. "Al Qaeda Propaganda Chief Killed in Strike: Officials," Agence France-Presse, November 1, 2008, http://afp.google.com/article/ALeqM5hFRj05wy4-L2HckvAXmlqSQKJ-yQ.
81. "Up to Fourteen Dead in US Missile Strike, Officials," Agence France-Presse, November 7, 2008, http://afp.google.com/article/ALeqM5jlbsctRKE5_0vBDyyTTr7nkLW0yg.
82. "US Drone Fires Missile into Pakistan, Up to 12 Dead," Reuters, November 14, 2008.
83. "US Drone Strike Kills 12 in Pakistan," *Times*, November 14, 2008; "US Drone Attack in Miranshah Kills 11," *Geo.TV*, November 14, 2008.
84. "US Drone Strike Kills 12 in Pakistan."
85. Bob Woodward, *Obama's Wars* (New York: Simon & Schuster, 2010), 26.
86. "Good Riddance," *Daily Times*, January 21, 2010.
87. Hassan Zaidi, "Army Chief Wanted More Drone Support," *Dawn*, May 20, 2011.
88. Ibid.

89. Tim Lister, "WikiLeaks: Pakistan Quietly Approved Drone Attacks, US Special Units," CNN, December 2, 2010, http://www.cnn.com/2010/US/12/01/wikileaks.pakistan.drones/index.html.

90. "US Drone Strikes Fail to Mobilize Pakistan Masses," *Dawn*, October 10, 2011.

91. Tayyab Siddiqui, "Pakistan's Drone Dilema," *Dawn*, July 18, 2010.

92. "Wikileaks: Gilani Open to Strikes on 'Right People,'" *Express Tribune*, December 1, 2010.

93. Gul, *Most Dangerous Place*, 206.

94. Karman Yousef, "Denying Wikileak Disclosure: Kayani Sought US Drone Support Only for Surveillance," *Express Tribune*, May 21, 2011.

95. *Los Angeles Times*, cited in "US Drones Back Up Pakistani Offensive," *Newser.com*, October 23, 2009, http://www.newser.com/story/72428/us-drones-back-up-pakistani-offensive.html.

96. "Wikileaks: Kayani Wanted More Drone Strikes in Pakistan," *Express Tribune*, May 20, 2011.

97. Gul, *Most Dangerous Place*, 216–17.

98. Matthew Rosenberg, Siobhan Gorman, and Jay Solomon, "Pakistan Lends Support for US Missile Strikes," *Wall Street Journal*, February 18, 2009.

99. Peter Bergen and Katherine Tiedemann, "The Drone War: Are Predators Our Best Option or Our Worst Enemy?" *New Republic*, June 3, 2009.

100. "US Officials Working to Sway Hearts and Minds in Islamabad," *Eurasia Insight*, July 8, 2009.

101. "US Drone Attack Kills Four Militants in Pakistan," ThaiIndian.com, November 19, 2008.

102. "Suspected US Missile Kills 3 in Pakistan," CNN, November 29, 2008, http://www.cnn.com/2008/WORLD/asiapcf/11/29/pakistan.missile/index.html.

103. "Transatlantic Plot (2006)," *New York Times*, July 8, 2010.

104. "US Kills Alleged Transatlantic Plot Leader, Report Says," *Guardian*, November 22, 2008.

105. "Suspected Missile Kills 3 in Pakistan."

106. "Suspected US Drone in Pakistan Kills 7," Reuters, December 12, 2008.

107. "US Drone Strike Kills Two in Waziristan," Geo.TV, December 15, 2008.

108. "At Least Eight Killed in Suspected Drone Attack in Pakistan," VOAnews.com, December 22, 2008; "At Least Eight Dead in Missile Strike, Officials," *Times of India*, December 22, 2008.

109. Peter Bergen and Katherine Tiedemann, *The Drone War* (Washington, DC: New America Foundation, June 3, 2009).

110. "Drone Attacks in Pakistan," *BuzzPK*, March 17, 2010, http://www.buzzpk.com/drone-attacks-in-pakistan/.

7. WHO IS BEING KILLED IN THE DRONE STRIKES?

1. Jeffrey Smith, Candace Rondeaux, and Joby Warrick, "2 US Airstrikes Offer Concrete Sign of Obama's Pakistan Policy," *Washington Post*, January 24, 2009.

2. Ibid.; "President Obama Orders Drone Attacks," *Times*, January 23, 2009.

3. Tara McKelvey, "Covering Obama's Secret War," *Columbia Journalism Review*, May–June 2011.

4. "Obama Is the Rambo of Drone Warfare," *Nation*, October 3, 2011.

5. Roggio and Mayer, "Charting the Data for US Airstrikes in Pakistan."

6. "Drones and the Man," *Economist*, July 30, 2011.

7. Entous, Gorman, and Barnes, "US Tightens Drone Rules."

8. Jessica Yellin, "Drone Program Something You 'Struggle With' Obama Says," *Political Ticker* (blog), CNN, September 10, 2012, http://politicalticker.blogs.cnn.com/2012/09/10/drone-program-something-you-struggle-with-obama-says/.

9. Gul, *Most Dangerous Place*, xxv.

10. Daniel Klaidman, *Kill or Capture: The War on Terror and the Soul of the Obama Presidency* (Boston: Houghton Mifflin Harcourt, 2012), 118.

11. Bill Adair, "In 2008 Obama Vowed to Kill Osama Bin Laden," *Tampa Bay Times*, May 1, 2011.

12. Joby Warrick, "Jan. 1 Attack by CIA Killed Two Leaders of Al Qaeda," *Washington Post*, January 9, 2009; "5 Killed in South Waziristan as US Drone Fired 3 Missiles in Karkot," *Nation*, January 27, 2009.

13. For a summary of all HVTs killed in the drone campaign, see Bill Roggio and Alexander Mayer's useful archive "Senior al Qaeda and Taliban Leaders Killed in US Airstrikes in Pakistan, 2004–2013," *Long War Journal*, http://www.longwarjournal.org/pakistan-strikes-hvts.php (accessed February 2013).

14. For an excellent account of the IMU, see Ahmed Rashid's book *Jihad: The Rise of Militant Islam in Central Asia* (New Haven, CT: Yale University Press, 2002).

15. Daniel Witter, *Uzbek Militancy in Pakistan's Tribal Region* (Washington, DC: Institute for the Study of War, January 27, 2011).

16. S. H. Khan, "US Strike Kills Eight Taliban in Pakistan, Officials," Agence France-Presse, August 27, 2009, http://www.google.com/hostednews/afp/article/ALeqM5gN_YaxdMssvvnU_CgoXAjjV-ZaMA.

17. "Frankfurt Airport Terror Plan 'Massive,'" *Herald Sun*, September 5, 2007.

18. "U.S.: Senior Al Qaeda Planner Likely Killed," CNN, December 11, 2009, http://www.cnn.com/2009/WORLD/asiapcf/12/11/pakistan.qaeda.operative/index.html.

19. Griff Witte, "U.S. Drone Strikes Kill 16 in Pakistani Region Thought to Be Al-Qaeda Home Base," *Washington Post*, December 18, 2009.

20. "Leading Taliban Leader Haji Omar Killed in Drone Strike," Free Library, December 31, 2009, http://www.thefreelibrary.com/Leading+Taliban+leader+Haji+Omar+killed+in+drone+attacks.-a0216247579.

21. Mary Louise Kelly, "Bin Laden Son Reported Killed in Pakistan," National Public Radio, July 22, 2009, http://www.npr.org/templates/story/story.php?storyId=106903109.

22. Greg Miller, "Increased Drone Strikes in Pakistan Killing Few High Value Militants," *Washington Post*, February 21, 2011.

23. Joby Warrick, "Attack by CIA Killed Two Leaders of Al Qaeda," *Washington Post*, January 9, 2009.

24. "Four Killed in Drone Attack in S. Waziristan," *Geo.TV*, January 2, 2009.

25. "Twenty Killed in US Drone Strike in S. Waziristan," *Geo.TV*, January 23, 2009, http://www.geo.tv/1-23-2009/33388.htm.

26. "President Obama Orders Drone Attacks," *Times*, January 23, 2009.

27. Pir Zubair Shah, "US Missile Strike Kills 25 Militants in Pakistan," Reuters, February 14, 2009; Pir Zubair Shah, "US Airstrike Kills 30 in Pakistan," *New York Times*, February 14, 2009.

28. "US Missile Attack Kills 30 in Kurram Agency," *Geo.TV*, February 16, 2009, http://www.geo.tv/2-16-2009/35249.htm.

29. Pir Zubair Shah, "US Missiles Hit Militants in Pakistan Killing Eight," *New York Times*, March 1, 2009.

30. "Suspected US Missile Kills 14 in Pakistan," Reuters, March 12, 2009.

31. "US Drone Kills Five in Pakistan," BBC News, March 16, 2009, http://news.bbc.co.uk/2/hi/south_asia/7945404.stm.

32. "Eight Killed in US Drone Attack in S. Waziristan," *Geo.TV*, March 25, 2009, http://www.geo.tv/3-25-2009/38186.htm; Bill Roggio, "US Airstrike Kills 8 in Baitullah Mehsud's Hometown," *Long War Journal*, March 25, 2009, http://www.longwarjournal.org/archives/2009/03/us_airstike_kills_8.php.

33. "US Drone Attack Kills 5 in S. Waziristan," *Geo.TV*, March 26, 2009, http://www.geo.tv/3-26-2009/38239.htm.

34. "Many Killed in US Drone Attack," BBC News, April 1, 2009, http://news.bbc.co.uk/2/hi/south_asia/7975871.stm.

35. "Suspected US Strike Kills 13 in Pakistan," Agence France-Presse, April 3, 2009; "13 Killed in US Drone Attack in Miramshah," *Geo.TV*, April 4, 2009, http://www.geo.tv/4-4-2009/39021.htm.

36. "US Drone Kills 3 in South Waziristan," *Geo.TV*, April 8, 2009, http://www.geo.tv/4-8-2009 /39356.htm; "Pakistan Reports 4 Dead in Suspected US Strike," VOA News, April 8, 2009.
37. "Three Suspected Dead in US Strike in Pakistan," Agence France-Presse, April 19, 2009; "Suspected US Drone Kills 3 in Pakistan," CNN, April 19, 2009, http://www.cnn.com /2009/WORLD/asiapcf/04/19/pakistan.drone/index.html.
38. "Suspected Missile Strike Kills 6 in Pakistan," Agence France-Presse, April 29, 2009.
39. "Scores Dead as Drone Hits Mehsud Stronghold," *Daily Times*, May 10, 2009.
40. "Drone Kills Eight in Pakistan," BBC News, May 12, 2009, http://news.bbc.co.uk/2/hi/south _asia/8045106.stm.
41. "Deaths in US Drone Raid in Pakistan," Al Jazeera, May 16, 2009; "Suspected Drone Missile Attack in Pakistan Kills 10," VOA News, May 16, 2009, http://www.voanews.com/content/a -13-2009-05-16-voa6-68643297/353852.html.
42. "13 Killed in Separate Attacks in Pakistan," CNN, June 14, 2009, http://www.cnn.com /2009/WORLD/asiapcf/06/14/pakistan.bomb/index.html.
43. "Air Raid Kills 9 in Pakistan," BBC News, June 18, 2009.
44. "Five Killed in S. Waziristan Drone Strike," *Geo.TV*, June 18, 2009.
45. "Suspected US Missile Strike Kills 17 in Pakistan," Agence France-Presse, June 22, 2009.
46. Ibid.; "US Missile Kills 51 in South Waziristan," *Daily Times*, June 24, 2009; "US Strikes Kill Dozens in Taliban Heartlands," Agence France-Presse, June 22, 2009; Pir Zubair Shah and Salman Masood, "US Drone Strike Said to Kill 60 in Pakistan," *New York Times*, June 23, 2009; Nic Robertson, "Taliban Defector Killed in Pakistan," CNN, June 23, 2009, http://www.cnn.com/2009/WORLD/asiapcf/06/23/pakistan.unrest/index.html; "US Drone Kills Scores of People in Pakistan Tribal Region," *Sunday Times*, June 25, 2009.
47. "13 Killed in US Drone Strikes in Pakistan," *Geo.TV*, July 3, 2009; Pir Zubair Shah and Ismail Khan, "Pakistan Army Crash Kills 26," *New York Times*, July 3, 2009.
48. "Deaths in Pakistan Missile Strike," Al Jazeera, July 8, 2009.
49. "Drones Kill Dozens in Pakistan," BBC News, July 8, 2009, http://news.bbc.co.uk/2/hi /8139739.stm.
50. "17 Killed as One More Drone Strike Hits SWA," *Geo.TV*, July 8, 2009.
51. "Drone Strike Kills 8 in S. Waziristan," *Daily Times*, July 11, 2009.
52. "US Missile Strike Kills 5 in Pakistan," CBC News, July 17, 2009, http://www.cbc.ca/news /world/story/2009/07/17/pakistan-bomb-blasts-oil-tankers.html.
53. "SWA Attack Kills 12 Extremists," *Geo.TV*, August 11, 2009.
54. Pir Zubair Shah and Lydia Polgreen, "US Drone Strike Kills 12 in Pakistan Border Region," *New York Times*, August 21, 2009.
55. "US Strike Kills Eight Taliban in Pakistan, Officials," Agence France-Presse, August 27, 2009, http://www.google.com/hostednews/afp/article/ALeqM5gN_YaxdMssvvnU _CgoXAjjV-ZaMA.
56. "Deaths in US Drone Raid," Al Jazeera, September 7, 2009; "Five Die in Pakistan Drone Strike," BBC News, September 8, 2009.
57. "Missile Strike in Northwest Pakistan Kills 10," VOA News, September 8, 2009.
58. "US Drone Attack Kills 4 in Pakistan, Officials," Agence France-Presse, September 13, 2009; "US Drone Attack Kills 8 in N. Waziristan," *Geo.TV*, September 14, 2009, http://www.geo.tv /9-14-2009/49185.htm.
59. "Drone Targets Militant Stronghold," BBC News, September 25, 2009, http://news.bbc.co.uk /2/hi/south_asia/8273997.stm.
60. Agence France-Presse, "Five Taliban Killed in US Strike in Pak," Free Library, September 29, 2009, http://www.thefreelibrary.com/Five+Taliban+killed+in+US+strike+in+Pak%3A +officials-a01612009960.
61. "Second Drone Attack Kills 7 in N. Waziristan," *Nation* (Pakistan), September 29, 2009.
62. Ibid.
63. "Suspected US Drone Kills 8 in Pakistan," Reuters, September 30, 2009.

64. Agence France-Presse, "US Strike on Suspected Militant Hideout in Pakistan," *Jakarta Globe*, October 15, 2009, http://www.thejakartaglobe.com/afp/us-strike-on-suspected -militant-hideout-in-pakistan/335649.

65. Hasbanullah Khan, "22 Killed as Drone Targets Taliban Shura Meeting in Bajaur," *Daily Times*, October 25, 2009.

66. "US Drone Kills Four in NWA," *Geo.TV*, November 5, 2009.

67. "US Drone Attack Kills 4 in NWA," *Geo.TV*, November 19, 2009, http://www.geo.tv/11-19 -2009/53188.htm.

68. "Pakistan Drone Attack Kills Eight Suspected Militants," BBC News, November 20, 2009, http://news.bbc.co.uk/2/hi/south_asia/8369795.stm.

69. "Suspected U.S. Missile Strike in Pakistan," Reuters, December 7, 2009.

70. "US Drone Strike Kills 3 in Miranshah," *Geo.TV*, December 8, 2010.

71. "15 Killed in Pakistan Drone Strikes," CNN, December 17, 2009, http://www.cnn.com /2009/WORLD/asiapcf/12/17/pakistan.drone/index.html.

72. "'Twelve Killed' in US Drone Strikes in Pakistan," BBC News, December 17, 2009, http://news.bbc.co.uk/2/hi/south_asia/8417756.stm.

73. "NWA Drone Strike Kills Eight," *Daily Times*, December 19, 2009.

74. "US Drone Strike Kills 13 in Saidgi Village: Security Officials," *Nation* (Pakistan), December 27, 2009.

75. "Missile Strike Kills 3 Militants in N Waziristan," *Geo.TV*, December 31, 2009, http://www.geo.tv /12-31-2009/55944.htm.

76. "US Censures Report on Drone Casualties," Al Jazeera, August 13, 2011, http://www.aljazeera .com/news/americas/2011/08/20118137276488676.html.

77. Scott Shane, "CIA Is Disputed on Civilian Toll in Drone Strikes," *New York Times*, August 11, 2011.

78. "Obama Administration Counter-Terrorism Strategy," C-Span, June 29, 2011, http://www.c -spanvideo.org/program/AdministrationCo.

79. Dan Robinson, "Obama Remark on Drone Strikes in Pakistan Stirs Controversy," VOA News, January 31, 2012, http://www.voanews.com/content/pakistan-repeats-condemnation -of-drone-strikes-138417439/151386.html.

80. Karen De Young, "After Remarks on Drones, White House Rebuffs Security Questions," *Washington Post*, January 31, 2012.

81. Shane, "CIA to Expand Use of Drones in Pakistan."

82. Ibid.

83. Tim Capaccio and Jeff Bliss, "US Said to Decrease Civilian Deaths after Increasing CIA Pakistan Strikes," Bloomberg, January 31, 2011.

84. Jeremy Scahill, "Georgetown Professor: Drones Are Not Killing Innocent Civilians in Pakistan," *Nation*, Mary 10, 2010.

85. Peter Bergen and Katherine Tiedemann, "The Hidden War," *Foreign Policy*, December 10, 2010.

86. Peter Bergen and Katherine Tiedemann, *The Year of the Drone: An Analysis of Drone Strikes in Pakistan, 2004–2011* (Washington, DC: New America Foundation, August 2, 2011), http://counterterrorism.newamerica.net/drones.

87. Peter Bergen and Jennifer Rowland, "Civilian Casualties Plummet in Drone Strikes," CNN, July 14, 2012, http://www.cnn.com/2012/07/13/opinion/bergen-civilian-casualties.

88. Eric Schmitt and Christopher Drew, "More Drone Attacks in Pakistan Planned," *New York Times*, April 6, 2009.

89. Shane, "CIA to Expand Use of Drones in Pakistan."

90. Cloud, "CIA Drones Have Broader List of Targets."

91. Capaccio and Bliss, "US Said to Reduce Civilian Deaths."

92. Peter Finn and Joby Warrick, "Under Panetta a More Aggressive CIA," *Washington Post*, March 21, 2010.

93. Sean Sullivan, "Panetta Defends Drone Strikes," *Washington Post*, February 3, 2013.

94. Mark Landler, "Civilian Deaths Due to Drones Are Not Many, Obama Says," *New York Times*, January 31, 2012.

95. Geoffrey Ingersoll, "Obama Finally Opens Up about the Massive Drone War," *Business Insider*, September 12, 2012.

96. Ken Dilanian, "Congress Zooms in on Drone Killings," *Los Angeles Times*, June 25, 2012, http://articles.latimes.com/2012/jun/25/nation/la-na-drone-oversight-20120625/2.

97. Greg Miller, "Brennan Defends Drone Strike Policy," *Washington Post*, February 7, 2013.

98. Scott Wilson, "Obama Nominates Chuck Hagel for Defense, John Brennan for CIA," *Washington Post*, January 27, 2013.

99. Andrew Lebovich, "Daily Brief: Officials Question Drone Strike Timing—Report," *AfPak Channel, Foreign Policy*, August 3, 2011, http://afpak.foreignpolicy.com/posts/2011 /08/03/daily_brief_officials_question_drone_strike_timing_report.

100. Tara Mckelevey, "Inside the Killing Machine," *Newsweek*, February 13, 2011.

101. Shane, "CIA Is Disputed."

102. Brian Mockenhaupt, "We've Seen the Future and Its Unmanned," *Esquire*, October 14, 2009.

103. Ken Dilanian, "CIA Drones May Be Avoiding Pakistani Civilians," *Los Angeles Times*, February 22, 2011.

104. Craig Whitlock and Greg Miller, "U.S. Covert Paramilitary Presence in Afghanistan Much Larger than Thought," *Washington Post*, September 22, 2011.

105. Robin Wright and Joby Warrick, "U.S. Steps Up Unilateral Strikes in Pakistan," *Washington Post*, March 27, 2008.

106. Woodward, *Obama's Wars*, 6.

107. Rosenberg, Gorman, and Solomon, "Pakistan Lends Support."

108. Nathan Hodge, "US Sharing Predator Video with Afghanistan, Pakistan," *Danger Room* (blog), *Wired*, November 19, 2008, http://www.wired.com/dangerroom/2008/11/in-a-presentati/.

109. Haq Nawaz Khan, "Rising Drone Attacks Hint at Strengthened CIA Spy Network in Waziristan," *Pakistan Today*, October 11, 2010.

110. "New Al Qaeda Book on 'Muslim Spies' Paints Picture of Weakened Group, Experts Say," Fox News, July 9, 2009, http://www.foxnews.com/story/0,2933,531161,00.html.

111. "Jihadist Website Posts Al Libi's 'Guidance on the Ruling of the Muslim Spy,'" Open Source Center, June 30, 2009, http://www.fas.org/irp/dni/osc/libi.pdf.

112. Ibid.

113. Adam Rawnsley, "CIA Drone Targeting Tech Revealed, Qaeda Claims," *Danger Room* (blog), *Wired*, July 8, 2009, http://www.wired.com/dangerroom/2009/07/infrared-beacons-guiding -cia-drone-strikes-qaeda-claims/.

114. Declan Walsh, "Mysterious 'Chip' Is CIA's Latest Weapon against Al-Qaida Targets Hiding in Pakistan's Tribal Belt," *Guardian*, May 31, 2009.

115. Rawnsley, "CIA Drone Targeting Tech Revealed."

116. Sami Yousafzai, "Predators on the Hunt in Pakistan," *Newsweek*, June 30, 2009.

117. "Al Qaeda Admits US Drone Attacks Working," CBS News, July 10, 2009.

118. Yousafzai, "Predators on the Hunt in Pakistan."

119. Bill Roggio, "Taliban Execute Four 'Spies' in Pakistan's Northwest," *Long War Journal*, February 5, 2011, http://www.longwarjournal.org/archives/2011/02/taliban_execute_4_sp.php.

120. Anthony Loyd, "US Drone Strikes in Pakistan Tribal Areas Boost Support for Taliban," *Times*, March 10, 2010.

121. Zia Khan, "Taliban Create Cell to Hunt 'Spies' Assisting US Drones," *Express Tribune*, March 28, 2011.

122. Zia Ur Rehman, "The Khurusan Mujahideen Seek to Eliminate Espionage in Waziristan," *Terrorism Monitor* 9, no. 13 (April 1, 2011): 3–5.

123. Bill Roggio, "Taliban Execute 4 Tribesmen 'Spying for America' in North Waziristan," *Long War Journal*, March 1, 2011, http://www.longwarjournal.org/archives/2011/03/taliban _execute_4_tr.php.

124. David Rohde, "The Drone War," *Reuters Magazine*, January 17, 2012.

125. Joby Warrick and Peter Finn, "Amid Outrage over Civilian Deaths in Pakistan, CIA Turns to Smaller Missiles," *Washington Post*, April 26, 2010.
126. Lockheed Martin, "Lockheed Martin's Scorpion Successful in Test Flight," news release, June 21, 2010, http://www.lockheedmartin.com/us/news/press-releases/2010/june /LockheedMartinsSCORPIONSu.html.
127. Steve Emerson, "In Defense of Drones: The ACLU Tries to Take Away an Effective and Humane Tool in the US Arsenal," *New York Post*, October 16, 2010.
128. "Drones in Pakistan: Out of the Blue," *Economist*, July 30, 2011.
129. Indo-Asian News Service, "Human Intelligence Aids Drone Attacks in Pakistan," India-Forums, December 30, 2010, http://www.india-forums.com/news/pakistan/301035 -human-intelligence-aid-drone-attacks-in-pakistan.htm.
130. "Ten More Killed in NWA Drone Attacks," PakTribune, March 18, 2010, http://paktribune .com/news/10-more-killed-in-NWA-drone-attacks-225651.html.
131. David Rohde and Kristen Mulvihill, *A Rope and a Prayer: A Kidnapping from Two Sides* (New York: Viking, 2010), 182, 239.
132. Salman Masood and Pir Zubair Shah, "CIA Drones Kill Civilians in Pakistan," *New York Times*, March 17, 2011.
133. Christine Fair, "Drones over Pakistan—Menace or Best Viable Option?" *The Blog* (blog), *Huffington Post*, August 2, 2010, http://www.huffingtonpost.com/c-christine-fair/drones -over-pakistan——m_b_666721.html.
134. Karen De Young, "US Officials Cite Gains against Al Qaeda in Pakistan," *Washington Post*, June 1, 2009.
135. Lister, "WikiLeaks: Pakistan Quietly Approved."
136. Walsh, "Mysterious 'Chip.'"
137. "Good Riddance."
138. Talat Masood, "Analysis: Time for a Rethink," *Daily Times*, November 27, 2008.
139. Zar ali Khan Musazai, "Drone Is an Anti-Terrorist Weapon, Mr. Nawaz Sharif," January 30, 2010, Global Hujra, http://www.khyberwatch.com/forums/showthread.php?8101-analysis -Drone-attacks-challenging-some-fabrications-Farhat-Taj&p=85459.
140. Fair, "Drones over Pakistan."
141. Farhat Taj, "Drone Attacks: Pakistan's Policy and the Tribesmen's Perspective," *Terrorism Monitor* 8, no. 10 (March 11, 2010): 8–10.
142. Nick Schifrin. "US Drones Strike with Deadly Accuracy," ABC News, November 19, 2008, http://abcnews.go.com/International/story?id=6289748&page=1.
143. Omar Waraich, "US Deaths in Pakistan Fuel Suspicion," *Time*, February 5, 2010.
144. "Pakistani Drone Victim Seeks to Put US on Trial," *Dawn*, December 23, 2010.
145. Farhat Taj, "Drone Attacks: Challenging Some Fabrications," *Daily Times*, January 2, 2010.
146. Zahir Sherzai, "Most of Those Killed in Drone Attacks Were Terrorists: Military," *Dawn*, March 9, 2011.
147. "Drone Attacks," *Dawn*, March 11, 2011, http://www.dawn.com/2011/03/11/drone-attacks -3.html.
148. "Pakistan: The State of the Nation," Al Jazeera, August 13, 2009, http://www.aljazeera .com/focus/2009/08/2009888238994769.html.
149. "60 Drone Hits Kill 14 Al-Qaeda Men, 687 Civilians," *News*, April 10, 2009.
150. "Over 700 Killed in 44 Drone Strikes in 2009," *Dawn*, January 2, 2010.
151. Brian Glyn Williams, Matthew Fricker, and Avery Plaw, "New Light on the Accuracy of the CIA's Predator Drone Campaign in Pakistan," *Terrorism Monitor* 8, no. 41 (November 11, 2010): 8–13.
152. Bergen and Tiedemann, *Year of the Drone*.
153. Roggio and Mayer, "Charting the Data for US Airstrikes in Pakistan."
154. Dexter Filkins, "Will Bin Laden's Death Alter U.S. Military Operations?" *Morning Edition*, National Public Radio, May 3, 2011, http://www.npr.org/2011/05/03/135941583/will-bin -laden-s-death-alter-u-s-military-operations.

155. Shahid Saeed, "Demystifying the Drone, Part I," *Daily Times*, October 27, 2010.

156. Taj, "Drone Attacks: Challenging Some Fabrications."

157. Bashir Ahmad Gwakh, "Drones, Pakistan Has No Leg to Stand On," Radio Free Europe/Radio Liberty, September 22, 2010, http://www.rferl.org/content/pakistan _commentary_drone_strikes/24336736.html.

158. David Kilcullen and Andrew Exum, "Death from Above, Outrage Down Below," *New York Times*, May 16, 2009.

159. Doyle McManus, "US Drone Attacks in Pakistan 'Backfiring' Congress Told," *Los Angeles Times*, May 3, 2009.

160. "Sami for Greater Alliance to Stop Drones," *Nation* (Pakistan), March 8, 2011.

161. "Drone Attacks in Pakistan," *BuzzPK*.

162. Tariq Saeed, "Drone Strikes Kill 6 in NWA," *Pakistan Observer*, March 12, 2011, http://pakobserver.net/detailnews.asp?id=80509.

163. Pakistan Body Count, http://www.pakistanbodycount.org/ (accessed December 2012).

164. Sharon Weinberger, "Pakistani Scholar Disputes US Drone Death Tallies," AOL News, May 19, 2010, http://www.aolnews.com/2010/05/19/pakistani-scholar-disputes-low-drone -death-tallies/.

165. "US Drones Killed 957 Civilians in 2010," *Irish Sun*, April 14, 2010.

166. Daniel Byman, *Do Targeted Killings Work?* (Washington, DC: Brookings Institution, July 14, 2009), http://www.brookings.edu/opinions/2009/0714_targeted_killings_byman.aspx.

167. "Cindy Sheehan: CIA Actions Killing Innocent People," YouTube video, 5:24, posted by Russia Today, January 15, 2010, http://www.youtube.com/watch?v=a3QumrijBIk.

168. Murray Wardrop, "Unmanned Drones Could Be Banned, Says Senior Judge," *Telegraph*, July 6, 2009.

169. Ted Rall, "US Drone Planes Have a Nearly Perfect Record of Failure," Common Dreams, January 18, 2006, http://www.commondreams.org/views06/0118-32.htm.

170. Brenda Norell, "Facebook Deletes Cindy Sheehan's Invite to CIA Drone Protest," *Narcosphere*, January 3, 2010, http://narcosphere.narconews.com/notebook/brenda-norrell /2010/01/facebook-deletes-cindy-sheehans-invite-cia-drone-protest.

8. SPIES, LAWYERS, TERRORISTS, AND SECRET BASES

1. Anwar Iqbal, "US Official Says Drones Using Pakistan Base," *Dawn*, February 14, 2009.

2. Greg Miller, "Feinstein Comment on U.S. Drones Likely to Embarrass Pakistan," *Los Angeles Times*, February 13, 2009.

3. "Pakistan Embassy in Washington Says No Foreign Bases in Country," Associated Press of Pakistan, February 15, 2009, http://www.app.com.pk/en_/index.php?option=com_content &task=view&id=68096&Itemid=1.

4. "Pak Bases Not Being Used Says Mukhtar," *Daily Times*, February 14, 2009.

5. Tom Coghlan, Zahid Hussain, and Jeremy Page, "Secrecy and Denial as Pakistan Lets CIA Use Airbase to Strike Militants," *Times*, February 17, 2009.

6. Iqbal, "US Official Says Drones Using Pakistan Base."

7. Shireen Mazari, "Pakistan's Complicity in Drone Attacks," *News*, February 15, 2009.

8. James Risen and Mark Mazzetti, "CIA Said to Use Outsiders to Put Bombs on Drones," *New York Times*, August 20, 2009.

9. Declan Walsh, "Pakistan Orders US out of Drone Base," *Guardian*, June 30, 2011.

10. "Pakistani Shooting Exposes Spy War," *Australian*, February 28, 2011.

11. "Probe Finds Connection between Davis, Drone Attacks," *Dawn*, February 18, 2011; Declan Walsh and Ewan MacAskill, "American Who Sparked Diplomatic Crises over Lahore Shooting Was CIA Spy," *Guardian*, February 20, 2011.

12. "Pakistani Shooting Exposes Spy War."

13. Jawad Awan, "Davis, CIA's Acting Chief in Pakistan," *Nation*, February 21, 2011.

14. Walsh and MacAskill, "American Who Sparked Diplomatic Crises."

15. Ibid.
16. "Munter Was Overruled on Halting Drone Strike," *Express Tribune*, August 3, 2011.
17. "Death Toll Reaches 44 in Drone Attack," AAJ News, March 18, 2011, http://www.aaj.tv /2011/03/death-toll-reaches-44-in-drone-attack/.
18. Salman Masood and Pir Zubair Shah, "CIA Drones Kill Civilians in Pakistan," *New York Times*, March 17, 2011.
19. "Pakistan, Calls for Revenge after Drone Kills 40," BBC, March 8, 2011, http://www.bbc.co.uk /news/world-south-asia-12784675.
20. Ibid.
21. "General Kayani Strongly Condemns Predator Strike," *Geo.TV*, March 17, 2011.
22. Associated Press, "US Pakistan Relations Increasingly Strained," *Dawn*, March 18, 2011, http://dawn.com/2011/03/18/us-pakistan-relationship-increasingly-strained/.
23. Munir Ahmed, "Pakistan Cancels US-Afghan Talks after Drone Strike," *Huffington Post*, March 18, 2011.
24. "US Intends to Step Up Drone Strikes," *New York Times*, April 9, 2009.
25. Gillmor, "Suicide Blast Kills 42 Pakistani Soldiers."
26. "Times Square Terrorism Suspect Angry over US Drones," MSNBC, May 10, 2010.
27. Aaron Katersky, "Faisal Shahzad Pleads Guilty in Times Square Car Bomb Plot, Warns of More Attacks," ABC, June 21, 2010, http://abcnews.go.com/Blotter/faisal-shahzad-pleads -guilty-times-square-car-bomb/story?id=10970094.
28. William Dalrymple, "Why Are the Taliban Winning in Afghanistan?" *New Statesman*, June 22, 2010.
29. Benjamin Weiser and Colin Moynihan, "A Guilty Plea in Plot to Bomb Times Square," *New York Times*, June 22, 2010.
30. "Pakistan Taliban Claim New York Bomb Attempt," Pkaffairs.com, http://www.pkaffairs.com /news.aspx?pageId=5278 (accessed December 2012).
31. Sami Yousufzai, "Pakistan Taliban Source: Times Square Bombing Attempt Was 'Revenge against America,'" *Newsweek*, May 10, 2010.
32. David Sanger, "US Pressure Helps Militants Focus Efforts," *New York Times*, May 7, 2010.
33. Anqoinette Crosby and Greg Miller, "CIA Shifts Focus to Killing Targets," *Washington Post*, August 30, 2011.
34. Zahid Hussain, "Attacker of CIA Is Linked to Taliban," *Wall Street Journal*, January 11, 2010.
35. "CIA Bomber Shown Vowing Revenge," BBC News, January 9, 2010, http://news.bbc.co.uk /2/hi/south_asia/8449875.stm.
36. Scott Shane and Eric Schmitt, "C.I.A. Deaths Prompt Surge in U.S. Drone Strikes," *New York Times*, January 22, 2010.
37. "Pakistan Tribesman to Sue CIA over Drone Deaths," Agence France-Presse, November 29, 2010.
38. Amir Mir, "A Revolving CIA Door in Pakistan," *Asia Times*, August 5, 2011.
39. Reza Sayah, "Pakistani Man Sues US over Drone Strikes," CNN, December 2, 2010, http://articles.cnn.com/2010-12-01/world/pakistan.drone.attack.lawsuit_1_drone-strike -drone-attacks-innocent-victims.
40. "Pakistani Role Suspected in Revealing US Spy's Name," *New York Times*, December 17, 2010.
41. Sayah, "Pakistani Man Sues US."
42. Code Pink, "San Diegans to Protest Predator Drone Makers and General Atomics CEO's Home," press release, May 17, 2010, http://www.codepink4peace.org/article.php?id=5418.
43. Cindy Sheehan, "First Protest against CIA Drone Attacks Coming to Langley, Virginia," *Cindy Sheehan's Soapbox* (blog), http://www.cindysheehanssoapbox.com/peacePage.html.
44. Farhan Bokhari, "Anti-Drone Protestors Blocked from Target, but Still Push Message," CBS News, October 8, 2012, http://www.cbsnews.com/8301-202_162-57527806/anti-drone -protesters-blocked-from-target-but-still-push-message/.

45. Tanver Ahmed, "Drone Attacks in Violation of International Laws," *Daily Times*, October 3, 2012.

46. Deborah Dupre, "Drones Kill Kids: Occupy DC Rights Defenders Shut Museum with Drone Exhibit," Examiner.com, October 8, 2011, http://www.examiner.com/article/drones-kill -kids-occupy-d-c-rights-defenders-shut-museum-with-drone-exhibit.

47. "Drone Protests Lead to Arrests at Tucson's Raytheon Missile Systems," Warisacrime.org, October 4, 2011, http://warisacrime.org/content/drone-protest-leads-arrest-tucsons -raytheon-missile-systems.

48. Ibid.

49. Dave Tobin, "37 People Who Protest at Hancock Air Base Near Syracuse against the Use of Drones Are Arrested on Friday," Syracuse.com, April 22, 2011, http://www.syracuse.com /news/index.ssf/2011/04/war_protestors_arrested_at_han.html.

50. "Drone Debate: Are Remotely Controlled Weapons Ethical," Syracuse.com, December 13, 2009, http://blog.syracuse.com/opinion/2009/12/drone_debate_are_remotely_cont.html.

51. Victoria Ross, "Use of Drones Is Both Immoral and Counterproductive," Buffalonews.com, November 2, 2010, http://www.highbeam.com/doc/1P2-26088616.html.

52. Denis Kucinich, "Drones Direct Hit upon Rule of Law," *Daily Kos*, August 17, 2011, http://www.dailykos.com/story/2011/08/17/1007855/-Drones-Direct-Hit-upon-Rule-of-Law.

53. Peter Singer, "Do Drones Undermine Democracy?" *New York Times*, January 21, 2012.

54. "Flight of the Drones," *Economist*, October 8, 2011.

55. Shane, "CIA to Expand Use of Drones in Pakistan."

56. Cassandra Vinograd, "UK Police Arrest 22 in Anti-Drone Demonstration," *Salon*, December 2, 2011.

57. "Occupy DC: No to Drone Strikes," *Dawn*, October 11, 2011.

58. Ian Harris, "Why Sci-Fi Weapons Don't Make Us Safer," *CounterPunch*, September 13, 2011, http://www.counterpunch.org/2011/09/13/drone-warfare/.

59. Ibid.

60. Edith M. Lederer, "UN Investigator Warns Use of Drones May Violate International Law," CNS News, October 28, 2009, http://cnsnews.com/news/article/un-investigator-warns-us -use-drones-may-violate-international-law.

61. "US Use of Drones Queried by UN," *New York Times*, October 27, 2009.

62. Harold Hongju Koh, "The Obama Administration and International Law" (speech, Annual Meeting of the American Society of International Law, Washington, DC, March 25, 2010), http://www.state.gov/s/l/releases/remarks/139119.htm.

63. "UN Special Rappateur Philip Alston Responds to US Defense of Drone Attacks' Legality," Democracy Now, April 10, 2010, http://www.democracynow.org/2010/4/1/drones.

64. "Defending Drones: The Laws of War and the Right to Self Defense," *Washington Post*, April 13, 2010, http://www.washingtonpost.com/wp-dyn/content/article/2010/04/12 /AR2010041204086.html.

65. James Meek, "Fort Hood Gunman Nadal Hasan 'Is a Hero,'" *Daily News*, November 9, 2009.

66. Lauren Frayer, "Times Square Suspect Reportedly Inspired by Radical Cleric," AOLNews.com, May 7, 2010, http://www.aolnews.com/2010/05/07/shahzad-reportedly -inspired-by-al-awlaki/.

67. Mark Mazzeti, Eric Schmitt, and Robert F. Worth, "American Born Al Qaeda Leader Is Killed by Missile in Yemen," *New York Times*, September 30, 2011, http://www.nytimes.com /2011/10/01/world/middleeast/anwar-al-awlaki-is-killed-in-yemen.html.

68. Carol Williams, "Awlaki's Death Rekindles Debate on Targeting Americans," *Los Angeles Times*, September 30, 2011.

69. "Awlaki Trained Suspected Christmas Jet Bomber How to Detonate Underwear, Document Reveals," Fox News, October 4, 2011, http://www.foxnews.com/us/2011/10/04/al-awlaki -trained-suspected-christmas-day-jet-bomber-how-to-detonate-underwear/; Ken Dilanian. "US Says Cleric Slain by Drone in Yemen Aided in Airliner Plot," *Los Angeles Times*, February 11, 2012.

70. Thomas Joscelyn, "Awlaki's Emails to Terror Plotter Show Operational Role," *Long War Journal*, March 2, 2011, http://www.longwarjournal.org/archives/2011/03/anwar_al _awlakis_ema.php.

71. Charlie Savage, "Secret Memo Made Legal Case to Kill a Citizen," *New York Times*, October 9, 2011.

72. "US Argument for Legal Drone Strikes Revealed," Al Jazeera, February 6, 2013, http://www .aljazeera.com/news/americas/2013/02/20132517311796860.html.

73. "Fiery American-Born Qaeda Militant Is Killed in Yemen by CIA Strike," *New York Times*, October 1, 2010.

74. Scott Shane, "US Approval of Killing of Cleric Causes Unease," *New York Times*, May 13, 2010.

75. Ibid.

76. "The Drone Wars: Weapons Like the Predator Kill Far Fewer Civilians," *Wall Street Journal*, January 9, 2010.

77. Emerson, "In Defense of Drones."

78. Terry Frieden, "Judge Throws Out Assassination Lawsuit," CNN, December 7, 2010, http://articles.cnn.com/2010-12-07/us/lawsuit.al.awlaki_1_case-on-procedural-grounds -bates-courts.

79. ACLU, "ACLU Seeks Information on Predator Drone Program," press release, March 16, 2010, http://www.aclu.org/national-security/aclu-seeks-information-predator-drone-program.

80. Dominic Rushe, Chris McGreal, Jason Burke, and Luke Harding, "Anwar al-Awlaki Death: US Keeps Role under Wraps to Manage Yemen Fallout," *Guardian*, September 30, 2011.

81. "Fiery American-Born Qaeda Militant."

82. Scott Shane, "Judging a Long Reach," *New York Times*, September 30, 2011.

83. Tim Funk, "Family of Al Qaeda Blogger Samir Khan 'Appalled' by US Actions," *McClatchey*, October 6, 2011, http://www.mcclatchydc.com/2011/10/06/v-print/126352/family-of-al -qaida-blogger-samir.html.

84. Sholto Byrnes, "It Was Wrong for the US to Kill Al Awlaki," *National*, October 8, 2011.

85. Carrie Dann and Casey Hunt, "With Bi-Partisan Aid, Paul Filibusters CIA Pick Brennan," NBC News, March 6, 2013, http://firstread.nbcnews.com/_news/2013/03/06/17210879 -with-bipartisan-aid-paul-filibusters-cia-pick-brennan.

86. Alexandra Petri, "Keep Talking Rand Paul, the Kentucky Senator's Blizzard Filibuster," *ComPost* (blog), *Washington Post*, March 6, 2013, http://www.washingtonpost.com/blogs /compost/wp/2013/03/06/keep-talking-rand-paul-the-kentucky-senators-blizzard -filibuster/.

87. Alec MacGillis, "Rand Paul Has a Point: The Time Has Come to Talk about Drone Strikes on US Soil," *New Republic*, March 6, 2013, http://www.newrepublic.com/article/112604 /rand-paul-drone-filibuster-senator-has-point#.

88. Kate Nocera, "Rand Paul Filibsuter Blasted by John McCain, Lindsey Graham." *Politico*, March 7, 2013, http://www.politico.com/story/2013/03/john-mccain-lindsey-graham-blast -rand-paul-filibuster-88564.html.

89. Richard Stevenson, "A Senator's Stance on Drones Scrambles Partisan Lines," *New York Times*, March 7, 2013, http://www.nytimes.com/2013/03/08/us/politics/mccain-and -graham-assail-paul-filibuster-over-drones.html.

90. Anthony Gregory, "Obama, the Ground Breaking President?" *Huffington Post*, October 7, 2011, http://www.huffingtonpost.com/anthony-gregory/obama-al-awlaki_b_998546.html.

91. David Cole, "Killing Our Citizens without Trial," *New York Review of Books*, November 24, 2011.

92. Byrnes, "It Was Wrong."

93. "Americans Say Government Was Right to Eliminate Islamist Militant," Angus-Reid.com, October 11, 2011, http://www.angus-reid.com/polls/44084/americans-say-government-was -right-to-eliminate-islamist-militant/.

94. Eric Holder, "Attorney General Eric Holder Speaks at Northwestern University School of Law" (speech, Northwestern University School of Law, Chicago, IL, March 5, 2012), http://www.justice.gov/iso/opa/ag/speeches/2012/ag-speech-1203051.html.

95. Ibid.

96. Tom Curry, "Holder Defends Targeted Killing Policy, Hints at Memo Release," *NBC Politics* (blog), NBC News, March 6, 2013, http://nbcpolitics.nbcnews.com/_news/2013/03/06 /17212054-holder-defends-targeted-killings-policy-hints-at-memo-release.

97. Bill Roggio, "AQAP Claims Media Emir Is Still Alive," *Threat Matrix* (blog), *Long War Journal*, October 29, 2011, http://www.longwarjournal.org/threat-matrix/archives/2011 /10/aqap_claims_media_emir_is_aliv.php.

98. Peter Finn and Greg Miller, "Anwar Awlaki's Family Speaks Out against His Son's Death in a Drone Strike," *Washington Post*, October 17, 2011.

99. Ken Dilanian, "Awlaki's Son, 16, Killed by Drone," *Sydney Morning Herald*, October 21, 2011.

100. Bill Roggio, "Anwar al Awlaki's Son Hoped to 'Attain Martyrdom as My Father Attained It,'" *Long War Journal*, December 8, 2011, http://www.longwarjournal.org/archives/2011 /12/anwar_al_awlakis_son.php.

101. Jo Becker and Scott Shane, "Secret 'Kill List' Proves a Test of Obama's Principles and Will," *New York Times*, May 29, 2012.

102. "Intelligence on Purported Europe Terror Plot Called 'Credible,'" NBC News, September 30, 2010, http://www.nbcnews.com/id/39413455/ns/world_news-europe/#.URqVb2ckf5Q.

103. Tristina Moore, "Terrorism Alert Focuses Spotlight on Germany," *Time*, October 12, 2010.

104. Troy McMullen and Anna Schecter, "German Terror Connection Grows: 45 More Suspects Being Tracked," ABC News, October 7, 2010, http://abcnews.go.com/Blotter/ahmed -siddiqui-german-terror-connection/story?id=11760048.

105. Richard Norton Taylor, "'Mumbai-Style' Terror Attack on UK, France and Germany Foiled," *Guardian*, September 29, 2010.

106. "Drone Killed British Taliban Plotter Reports Say," *Guardian*, October 6, 2010.

107. "British Man Set to Lead British Terror 'Army' Killed in Pakistan Drone Strike against Militants," *Daily Mail*, October 6, 2010.

108. "Pakistan Drone Attack Kills Germans in Response to Europe Terror Plot," *Christian Science Monitor*, October 5, 2010.

109. "Drone Attack Kills Two Britons in Pakistan," BBC News, December 16, 2010, http://www.bbc .co.uk/news/uk-12006061.

110. Pam Benson et al., "Intelligence, Potential Plots Are Factors in Drone Attack Increase," CNN, September 28, 2010.

111. John Reed, "RQ-170 Stealth Drone Used in Bin Laden Raid," *Defense Tech*, May 18, 2011, http://defensetech.org/2011/05/18/rq-170-sentinel-stealth-drone-used-in-bin-laden-raid/.

112. Greg Miller, "CIA Flew Stealth Drones into Pakistan to Monitor Bin Laden House," *Washington Post*, May 17, 2010.

113. Gloria Borger, "Attack on Osama Bin Laden Was Years in the Making," CNN, May 4, 2011, http://www.cnn.com/2011/OPINION/05/02/borger.ksm.bin.laden/index.html.

114. For an in depth account of the raid see Williams, *Afghanistan Declassified*.

115. David Sanger and Eric Schmitt, "As Rift Deepens, Kerry Has a Warning for Pakistan," *New York Times*, May 14, 2011.

116. Christina Lamb, "Killing of Al Qaeda Leaders Hurting," *Australian*, September 19, 2011.

117. Sanger and Schmitt, "As Rift Deepens."

118. Singer, "Do Drones Undermine Democracy?"

119. John Pike quoted in Sam Biddle, "Drones Mean the Iraq War Is Never Over," *Gizmodo*, October 21, 2011, http://gizmodo.com/5852228/drones-mean-the-iraq-war-is-never-over.

120. "US Drones Bombed Libya More than Pakistan," *RT*, October 20, 2011, http://rt.com/usa /news/us-libya-pakistan-drone-325/.

121. Spencer Ackerman, "Libya: The Real US Drone War," *Danger Room* (blog), *Wired*, October 20, 2011, http://www.wired.com/dangerroom/2011/10/predator-libya/.

122. Thomas Harding, "Colonel Gaddafi Killed: Convoy Bombed by Drone Flown by Pilot in Las Vegas," *Telegraph*, October 20, 2011.

123. Kristina Wong, "US Drone a Tech Challenge for Iran," *Washington Times*, January 12, 2012.

124. Scott Peterson, "Iran Hacked US Drone Says Iranian Engineer," *Christian Science Monitor*, December 15, 2011.

125. Mark Clayton, "Did Iran Hijack the 'Beast'? US Experts Cautious about Bold Claims," *Christian Science Monitor*, December 16, 2011.

126. Martha Raddatz and Luis Martinez, "Is US Drone Shown on Iran TV Real?" ABC News, December 8, 2011, http://abcnews.go.com/Blotter/us-rq-170-sentinel-stealth-drone-shown -iran/story?id=15115781.

127. Scott Shane and David Sanger, "Drone Crash in Iran Reveals Secret US Drone Effort," *New York Times*, December 7, 2011.

128. Some claimed that the drone shown on Iranian television was a fake, but this has not been ascertained.

129. "Iran Airs Footage of Downed US Drone," *Press TV*, December 8, 2011.

130. Lee Ferran, "Obama: Hey Iran, Can We Get Our Drone Back?" ABC News, December 12, 2011, http://abcnews.go.com/Blotter/obama-asks-iran-rq-170-sentinel-drone-back/story ?id=15140133.

131. Dave Majumdar, "Iran's Captured RQ 170. How Bad Is the Damage?" *Air Force Times*, December 9, 2011; "Iran Displays RQ 170 Sentinel," *Global Military Review*, December 2011.

132. Roggio and Mayer, "Charting the Data for US Airstrikes in Pakistan."

133. Entous, Gorman, and Barnes, "US Tightens Drone Rules."

134. Ibid.

135. Barbara Starr, "US Pauses Drone Strikes in Pakistan," CNN, December 24, 2011, http://www .cnn.com/2011/12/24/world/asia/pakistan-us-drones.

136. U.S. Department of Defense, "Department of Defense Statement Regarding Investigation Results into Pakistan Cross-Border Incident," news release, December 22, 2011, http://www .defense.gov/releases/release.aspx?releaseid=14976.

137. Ibid.

138. Faisal Aziz, "US Leaving Drone Base Won't Have Big Impact on Air War," Reuters, December 12, 2011.

139. "Pakistan to Down Intruding US Drones," *Press TV*, December 11, 2011, http://www.presstv.ir /detail/215014.html.

140. "18 Killed in Suspected Drone Attacks in Pakistan," CNN, November 16, 2011, http://www.cnn .com/2011/11/16/world/asia/pakistan-drone-attacks.

141. Bill Roggio, "US Will Break Lull in Drone Strikes Only to Hit a High Value Target: Pakistani Official," *Long War Journal*, December 20, 2011, http://www.longwarjournal.org/threat -matrix/archives/2011/12/us_will_break_lull_in_drone_st.php.

142. "To Pacify Pakistan US Asked for Permission for Jan. Drone Strike," *Express Tribune*, January 17, 2012.

143. Eric Schmitt, "Lull in US Drone Strikes Aids Militants in Pakistan" *New York Times*, January 6, 2012.

144. Haji Mujtaba, "Suspected US Drone Kills Four Militants in Pakistan," Reuters, January 10, 2012.

145. Rasool Dawar, "US Fires First Drone in 6 Weeks, 4 Dead," *Guardian*, January 11, 2012.

146. Ibid.

147. Scott Shane, "Drone Strike Kills Qaeda Operative in Pakistan, US Says," *New York Times*, January 20, 2012.

148. "Hakimullah Not Dead: Pakistani Taliban," *Times of India*, January 16, 2012.

149. Chris Allbritton, "Exclusive: How Pakistan Helps the US Drone Campaign," Reuters, January 22, 2010.

150. Declan Walsh and Eric Schmitt, "Abu Yahya al Libi of Al Qaeda Said to Be Killed," *New York Times*, June 5, 2012.

151. Associated Press, "Report: US Drone Kills Somali Al Qaeda Insurgent," Fox News, January 22, 2012, http://www.foxnews.com/world/2012/01/22/report-us-drone-strike-kills-somali -al-qaeda-insurgent/.

152. Ian Cobain, "British Al Qaida Member Killed in Drone Attack in Somalia," *Guardian*, January 22, 2012.

153. Julian Barnes, "US Expands Drone Flights to Take Aim at East Africa," *Wall Street Journal*, September 21, 2011; Jim Lobe, "US Expanding Network of Drone Bases to Hit Somalia, Yemen," Inter Press Service, September 21, 2011, http://www.ipsnews.net/2011/09/us -expanding-network-of-drone-bases-to-hit-somalia-yemen/; "Who Will the Drones Target?" Associated Press, May 22, 2012.

154. Saleh ali Saleh Nabhan was killed by JSOC in September 2009 and Aden Hashi Farah "Ayrow" was killed in a May 2008 air strike.

155. Greg Jaffe and Karen De Young, "US Drone Strike Targets Two Leaders of Somali Group Allied to Al Qaeda, Official Says," *Washington Post*, June 29, 2011; Tristan McConell, "Drone Wars: Somalia Becomes the Latest Front," *GlobalPost*, October 10, 2011, http://www .globalpost.com/dispatch/news/regions/africa/111007/drone-wars-somalia-al-shabaab-cia.

156. "US Drones Strike al Shabab in Somalia," *GlobalPost*, September 26, 2011, http://www .globalpost.com/dispatches/globalpost-blogs/africa-emerges/drone-attack.

157. Emma Slater and Chris Woods, "Iranian TV Accused of Faking Reports of Somali Drone Strikes," *Guardian*, December 2, 2011; "US Drone Strike Kills 35 in S. Somalia," *PressTV*, September 6, 2011, http://presstv.com/detail/197622.html.

158. David Axe, "Hidden History: America's Secret Drone War in Africa," *Danger Room* (blog), *Wired*, August 13, 2012, http://www.wired.com/dangerroom/2012/08/somalia-drones/.

159. "Profile: Al Qaeda in the Arabian Peninsula," Al Jazeera, May 9, 2012.

160. Bill Roggio and Bob Barry, "Charting the Data for US Air Strikes in Yemen, 2002–2013," *Long War Journal*, http://www.longwarjournal.org/multimedia/Yemen/code/Yemen-strike.php (accessed March 8, 2013).

161. "Air Fighter Blast Al Qaeda's Weapons Stock," *Yemen Post*, May 23, 2012.

162. "Yemen Says Airstrikes, Troops Kill Seven Militants," Reuters, May 19, 2012, http://www.reuters .com/article/2012/05/19/us-yemen-idUSBRE84I07V20120519.

163. "11 Militants in Fiercest Battle to Retake Key Areas in Yemen," *Yemen Post*, June 11, 2012.

164. Peter Bergen and Jennifer Rowland, "Obama Ramps Up Covert War in Yemen," CNN, June 11, 2012, http://www.cnn.com/2012/06/11/opinion/bergen-yemen-drone-war.

165. Jacob Zenn, "US Drones Circle over Philippines," *Asia Times*, February 29, 2012.

166. "Abu Sayyaf and Jemaah Islamiya Members Killed in Airstrike," *Jamestown Foundation Briefs* 3, no. 2 (February 29, 2012), http://mlm.jamestown.org/single/?tx_ttnews[tt_news] =39067&tx_ttnews[backPid]=546&cHash=64e28b4fd0e8afdf68a47917ee6f1457.

167. Mark Mazzetti, "The Drone Zone," *New York Times*, July 6, 2012.

168. Chris Woods and Christina Lamb, "Obama Terror Drones: CIA Tactics in Pakistan Include Targeting Rescuers and Funerals," Bureau of Investigative Journalism, February 4, 2012, http://www.thebureauinvestigates.com/2012/02/04/obama-terror-drones-cia-tactics-in -pakistan-include-targeting-rescuers-and-funerals/

169. Glenn Greenwald, "US Said to Target Rescuers at Drone Sites," *New York Times*, February 5, 2012.

170. Ibid.

171. Woods and Lamb, "Obama Terror Drones."

172. "Blast Kills at Least Twenty in Pakistan," *Washington Post*, June 20, 2007.

173. Bill Roggio, "US Drones Kill 10 'Militants' in North Waziristan Strike," *Long War Journal*, February 8, 2012, http://www.longwarjournal.org/archives/2012/02/us_drones_killed _8_m.php.

174. Ismail Khan and Declan Walsh, "Drone Kills Pakistani Militant, Officials Say," *New York Times*, February 9, 2012.

175. Bill Roggio, "Top North Waziristan Taliban Leader Bahadar Rumored Killed in US Strike," *Long War Journal*, March 11, 2010, http://www.longwarjournal.org/archives/2010/03/top_north _waziristan.php.

176. "US Drone 'Kills 10' in Pakistan," *BBC News*, February 8, 2012, http://www.bbc.co.uk /news/world-asia-16939853.

177. Taj, "Drone Attacks: Challenging Some Fabrications."

178. "Pakistan Drone Attack 'Kills' Top Militants," *BBC News*, October 27, 2011.

179. "Drone Attacks, A Survey," *News*, March 5, 2009.

180. "Cables Show US Special Operations in Pakistan," Al Jazeera, May 21, 2011; "Leaked US Diplomatic Messages Show Pakistan Military Requested More Drone Strikes and Help from US Special Forces," Al Jazeera, May 21, 2011.

181. Farhat Taj, "Drone Attacks and US Reputation," *Daily Times*, February 6, 2010.

182. "US Drone Kills Scores of People in Tribal Zone," *Times*, June 25, 2009.

183. Haq Nawaz Khan and Karin Brulliard, "Bombing at Funeral Kills at Least 25 in Northwestern Pakistan," *Washington Post*, September 15, 2011; Associated Press, "Pakistan Suicide Bombing Kills 36 at Funeral," CBC News, March 9, 2011, http://www.cbc.ca /news/world/story/2011/03/09/pakistan-suicide-bombing.html.

184. Taj, "Drone Attacks: Challenging Some Fabrications."

9. THE ARGUMENT FOR DRONES

1. "US Air Strikes in Pakistan Called Very Effective," CNN, May 18, 2009, http://www.cnn.com /2009/POLITICS/05/18/cia.pakistan.airstrikes/.

2. Shane and Schmitt, "C.I.A. Deaths Prompt Surge."

3. John Cushman, "Bin Laden Plot against Obama Laid Out in Documents," *New York Times*, March 16, 2012.

4. Ravi Somaiya, "Drone Strike Prompts Suit, Raising Fears for U.S. Allies," *New York Times*, January 30, 2013, http://www.nytimes.com/2013/01/31/world/drone-strike-lawsuit-raises -concerns-on-intelligence-sharing.html.

5. Brian Glyn Williams, "Mullah Omar's Missiles: A Field Report on Suicide Bombers in Afghanistan," *Middle East Policy* 15, no. 4 (Winter 2008).

6. Syed Shazad, "Drones Ever Closer to Pakistan's Militants," *Asia Times*, October 27, 2011.

7. Bill Roggio, "Taliban Rebuild Children's Suicide Camp in South Waziristan," *Long War Journal*, October 6, 2008, http://www.longwarjournal.org/archives/2008/10/taliban _rebuild_chil.php.

8. Sayed Bukhari, "Misplaced Sense of Sovereignty," letter to the editor, *Dawn*, September 27, 2008.

9. Ajmal, comment on Nadeem Paracha, "Drone Attacks: The Truth Is Out," *Café Black* (blog), *Dawn*, April 28, 2011, http://www.dawn.com/2011/04/28/drone-attacks-the-truth-is -out.html.

10. Ifran Husain, "Howling at the Moon," *Dawn*, January 9, 2010.

11. Farhan Bokhari, "Transcript: Interview with Ali Asif Zardari," *Financial Times*, September 15, 2009, http://www.ft.com/cms/s/0/36a57efa-a205-11de-81a6-00144feabdc0.html #axzz2N5DyPlGL.

12. Mark Mazzetti and Eric Schmitt, "Shaky Pakistan Is Seen as Target of Al Qaeda Plots," *New York Times*, May 10, 2009.

13. "TTP's Qari Hussain Killed in Drone Attack," *Tribune Express*, October 15, 2010.

14. Daniel Byman, "Taliban vs. the Predator: Are Targeted Killings in Pakistan a Good Idea?" *Foreign Affairs*, March 18, 2009.

15. Masood and Shah, "CIA Drones Kill Civilians in Pakistan."

16. Rohde, "Drone War."

17. Mockenhaupt, "We've Seen the Future."

18. "Pakistan Taliban Member Describes Counter Measures against UAV Attacks," *Terrorism Monitor* 7, no. 31 (October 23, 2009): 3–5.

19. Karen Bruillard and Haq Nawaz Khan, "Pakistanis Tie Slayings to Surge in US Strikes," *Washington Post*, December 24, 2010.

20. Indo-Asian News Service, "Human Intelligence Aids Drone Attacks in Pakistan," India-Forums, December 30, 2010, http://www.india-forums.com/news/pakistan/301035-human -intelligence-aid-drone-attacks-in-pakistan.htm.

21. Alex Rodriguez and David Zucchino, "US Drone Attacks in Pakistan Get Mixed Response," *Los Angeles Times*, May 2, 2010.

22. Walsh, "Mysterious 'Chip.'"

23. Ibid.

24. Jane Perlez and Pir Zubair Shah, "Drones Batter Al Qaeda and Its Allies in Pakistan," *New York Times*, April 4, 2010.

25. Munir Ahmad, *The Battle for Pakistan: Militancy and Conflict in Balochistan*, National Security Studies Program Policy Paper (Washington, DC: New America Foundation, September 2011).

26. Eric Schmitt and Jane Perlez, "Strikes Worsen Qaeda Threat Pakistan Says," *New York Times*, February 24, 2009.

27. "Al Qaeda under Pressure: US Drones Have Al Qaeda on the Run," CBS News, July 10, 2009, http://www.cbsnews.com/8301-18563_162-5151547.html.

28. Mayer, "Predator War."

29. Katherine Tiedemann, "Drone Attacks in Tribal Pakistan Force Al Qaeda into Urban Areas," *Times*, August 8, 2009; Imtiaz Ali, "Karachi Becoming a Taliban Safe Haven?" *Combating Terrorism Center Sentinel* 3, no. 1 (January 2010): 17–20.

30. Associated Press, "Jihadist Shift Seen in Pakistan, Fewer Arabs," Fox News, September 3, 2012, http://www.foxnews.com/world/2012/09/03/jihadist-shift-seen-in-pakistan-fewer-arabs/.

31. Tom Gjelten, "U.S. Officials: Al-Qaida Leadership Cadre 'Decimated,'" *Morning Edition*, National Public Radio, February 3, 2009, http://www.npr.org/templates/story/story.php ?storyId=100160836.

32. Peter Taylor, "Drones 'Winning' War against Al Qaeda Says Ex-CIA Head," BBC News, March 20, 2011, http://www.bbc.co.uk/news/world-12784129.

33. Emerson, "In Defense of Drones."

34. Ali Chishti, "Al Qaeda Shifting to Pakistan's Urban Areas," *Daily Times*, August 25, 2010.

35. David Ignatius, "What Would Reconciliation Look Like for the US and Taliban?" *Washington Post*, June 29, 2010.

36. Associated Press, "Al Qaeda Leader Admits Facing Pressure from Drones," Fox News, January 27, 2011, http://www.foxnews.com/world/2011/01/27/al-qaida-leader-admits -facing-pressure-drones.

37. Greg Miller, "US Predator Strikes Causing 'Confusion' in Al Qaeda Ranks," *Los Angeles Times*, March 22, 2009.

38. Greg Miller, "Predator Strikes in Pakistan: U.S. Says Drones Ravage Al Qaeda," *Chicago Tribune*, March 22, 2009.

39. "Taliban Dismisses US Claims about Death of Osama's Son," *Hindu*, July 4, 2009.

40. Carol Grisanti and Mushtaq Yusufzai, "Taliban Style Justice for Alleged US Spies," *World Blog*, NBC News, April 17, 2009, http://worldblog.nbcnews.com/_news/2009/04/17 /4376383-taliban-style-justice-for-alleged-us-spies.

41. Mark Macdonald, "Are Drone Strikes Worth the Cost?" *New York Times*, August 22, 2012.

42. Christopher Swift, "The Drone Blowback Fallacy: Strikes in Yemen Aren't Pushing People to Al Qaeda," *Foreign Affairs*, July 1, 2012.

43. Karen De Young and Julie Tate, "Increased Drone Strikes in Pakistan Killing Few High Ranking Militants," *Washington Post*, February 20, 2011.

44. "Bin Laden Was Trying to Build Drone-Decimated Network," *Japan Times*, July 3, 2011.

45. Greg Miller, "Bin Laden Document Trove Reveals Strain on Al Qaeda," *Washington Post*, July 1, 2011.

46. Spencer Ackerman, "White House: Al Qaeda Is Toast (as Long as These Shadow Wars Last Forever)," *Danger Room* (blog), *Wired*, June 29, 2011, http://www.wired.com/dangerroom /2011/06/white-house-al-qaida-is-toast/.

47. Mark Mazzetti, "Al Qaeda Affiliates Growing Independent," *New York Times*, August 29, 2011.
48. "The Drone Wars," *Wall Street Journal*, January 9, 2010.
49. "Predators and Civilians," *Wall Street Journal*, July 14, 2009.
50. "The Ethics of Unmanned Warfare," United Press International, December 8, 2010.
51. William Saletan, "In Defense of Drones: They're the Worst Form of War, Except for All the Others," *Slate*, February 19, 2013, http://www.slate.com/articles/health_and_science/human _nature/2013/02/drones_war_and_civilian_casualties_how_unmanned_aircraft_reduce _collateral.html.
52. Greg Miller, "Under Obama a Huge Rise in Drone Strikes," *Washington Post*, January 1, 2012.
53. Noah Shachtman, "How the Afghanistan Air War Got Stuck in the Sky," *Wired*, December 8, 2009, http://www.wired.com/magazine/2009/12/ff_end_air_war/.
54. Christine Fair, "The Obama Administration Won't Tell the Truth about America's New Favorite Weapon—But That Doesn't Mean the Critics are Right," *Foreign Policy*, May 28, 2010.
55. "Flight of the Drones."
56. Scott Shane, "Yemen's Leader Praises US Drone Strikes," *New York Times*, September 19, 2012; Greg Miller, "In Interview Yemeni President Acknowledges Approving US Drone Strikes," *Washington Post*, September 29, 2012.
57. Graeme Smith, "Pakistan's Deadly Robots in the Sky," *Globe and Mail*, October 10, 2010.
58. Sarmad Tariq, "Drone Strikes Kill Militants, Others Kill the Innocent," *Welcome to Pakistan* (blog), *Express Tribune*, January 18, 2013, http://blogs.tribune.com.pk/story/15697 /drone-strikes-kill-militants-others-kill-the-innocent/.
59. John Schmidt, *The Unraveling: Pakistan in the Age of Jihad* (New York: Farrar, Strauss and Giroux, 2011), 155–56, 158, 175.
60. Taj, "Drone Attacks: Challenging Some Fabrications."
61. Taj, "Drone Attacks: Pakistan's Policy."
62. "Mapping US Drone and Islamic Militant Attacks in Pakistan," BBC News, July 22, 2010, http://www.bbc.co.uk/news/world-south-asia-10648909.
63. Taj, "Drone Attacks: Challenging Some Fabrications."
64. Azizullah Khan Khetran, "In Favor of Drone Attacks," *Daily Times*, November 6, 2010.
65. Charles Recknagel, "Drone Attacks Stoke Debate in Pakistan," EurasiaNet.org, October 17, 2010, http://www.eurasianet.org/node/62172.
66. Taj, "Drone Attacks: Challenging Some Fabrications."
67. Hasan Zaidi, "Army Chief Wanted More Drone Support," *Dawn*, May 20, 2011.
68. Sayeda Bukhari, "Are the Drone Attacks Not a Violation of Our Sovereignty?" trans. Zain Jamshaid, *Watching America*, October 18, 2011, http://watchingamerica.com/News /127159/are-the-drone-attacks-not-a-violation-of-our-sovereignty/.
69. Schmidt, *Unraveling*, 164, 175.
70. Peter Bergen, *The Longest War: The Enduring Conflict between America and Al-Qaeda* (New York: Free Press, 2011), 264.
71. David Ignatius, "Drones Allow CIA, Pakistan to Get Revenge," *New Haven Register*, February 6, 2010, http://www.nhregister.com/articles/2010/02/06/opinion /doc4b6ce4afc7687022423909.txt.
72. Anthony Loyd, "US Drone Strikes in Pakistan Tribal Area Boost Support for Taleban," *Times*, January 2, 2010.
73. Gwakh, "Drones, Pakistan Has No Leg to Stand On."
74. Azizullah Khan Khetran, "In Favour of Drone Attacks," *Daily Times*, November 6, 2010.
75. Taj, "Drone Attacks: Pakistan's Policy."
76. Farhat Taj, "Drone Attacks: A Survey," *News*, March 5, 2009.
77. Raza Habib, "Why I Support Drone Attacks," *Chowk.com*, July 26, 2010.
78. Khetran, "In Favour of Drone Attacks."

79. Pervez Hoodbhoy, "Drones: Theirs and Ours," *Viewpoint*, 2010, http://www.viewpointonline .net/drones-theirs-and-ours.html.

80. Aamir Latif, "Pakistani Tribes Caught between Taliban and US Airstrikes," *U.S. News & World Report*, February 11, 2009.

81. "Logic of Drone Attacks," *Daily Times*, April 3, 2009.

82. Taj, "Drone Attacks: A Survey."

83. Aryana Institute for Regional Research and Advocacy, "Peshawar Declaration," December 12, 2009, www.airra.org/newsandanalysis/Peshawardeclaration.php.

84. Farhat Taj, "Pakistani Responses to the CIA's Predator Drone Campaign against the Taliban and Al Qaeda," *Terrorism Monitor* 8, no. 7 (February 19, 2010).

85. Farhat Taj, "Drone Attacks. A Critical Perspective," *Viewpoint*, no. 141, http://www .viewpointonline.net/drone-attacks-a-critical-perspective.html.

86. Comment on Pakhtunkhwa Peace Forum Facebook page, February 20, 2010, https://www.facebook.com/Pashtoonwali/posts.

87. "Petition: Resume Drone Strikes on Taliban, Al Qaeda and Sipah e Sahba," petition posted by "a global citizen," *Change.org*, January 8, 2012, http://www.change.org/petitions/petition -resume-drone-attacks-on-taliban-al-qaeda-and-sipah-e-sahaba.

88. Schifrin, "US Drone Strikes with Deadly Accuracy."

89. Nadeem Sarwar, "US Drone Brings Torment, Hope in Pakistan," Deutsche Presse Agentur, March 19, 2010.

90. Taj, "Drone Attacks: Pakistan's Policy."

91. Ibid.

92. Ibid.

93. Fair, "Drones over Pakistan."

94. "Capturing the Punjabi Imagination: Drones and the 'Noble Savage,'" Reuters, November 13, 2011.

95. Salman Masood, "Pakistani General in Twist, Credits Drone Strikes," *New York Times*, March 9, 2011.

96. Taj, "Drone Attacks: Challenging Some Fabrications."

97. Taj, "Drone Attacks and US Reputation."

98. Farhat Taj, "Drone Attacks: A Critical Perspective," *Viewpoint*, 2010, http://www .viewpointonline.net/drone-attacks-a-critical-perspective.html.

99. "European Jihadis Flock to Pakistan for Terrorist Training," *Pakistan Conflict Monitor*, August 3, 2010.

100. "Anti-Taliban Lashkar in Pakistan?" *Daily Times*, June 18, 2011.

101. Matt Dupee, "Anti-Taliban Tribal Leader Killed in Northwest Pakistan," *Long War Journal*, December 2, 2011, http://www.longwarjournal.org/archives/2011/12/anti-taliban _tribal.php.

102. Khan, *Armageddon in Pakistan: The Crisis of a Failed Feudal Economy* (Pittsburgh: Rose Dog Books, 2009), 178.

103. Ibid.

104. Swift, "Drone Blowback Fallacy."

105. Amitai Etzioni, "The Great Drone Debate," *Military Review*, March–April 2013.

10. THE ARGUMENT AGAINST DRONES

1. "Pakistanis Blast US Air Raids," *Star*, October 31, 2009.

2. Ayaz Khan, "Drone Attacks Continue," *Pakistan Observer*, March 10, 2009.

3. Pew Research Global Attitudes Project, *Concern about Extremist Threat Slips in Pakistan* (Washington, DC: Pew Research Center, July 29, 2010).

4. Declan Walsh, "US Extends Drone Strikes to Somalia," *Guardian*, June 30, 2011.

5. Los Angeles Times, "More Pakistanis See US as Enemy, Despite Aid," *SFGate*, June 28, 2012, http://www.sfgate.com/world/article/More-Pakistanis-see-U-S-as-enemy-despite-aid -3672132.php.

6. New American Foundation, *FATA: Inside Pakistan's Tribal Regions*, http://pakistansurvey.org/ (accessed December 2012).

7. Jane Perlez, "Petraeus, in Pakistan, Hears Complaints about Missile Strikes," *New York Times*, November 3, 2008.

8. Ibid.

9. "Bigger Role for US CIA Drones in Pakistan," *Australian*, October 28, 2008.

10. "SWA Action to Conclude Well before Time: PM Gilani," Pak Tribune, November 16, 2009. http://paktribune.com/news/SWA-action-to-conclude-well-before-time-PM-Gilani-221378.html.

11. Hasnain Kazim, "How the CIA Uses Pakistan as a Launch Pad for Drones," *Der Speiegel*, March 12, 2010.

12. Liam Stack, "Fresh Drone Attacks in Pakistan Reignite Debate," *Christian Science Monitor*, July 8, 2009.

13. Saed Shah, "Anger in Pakistan at US Plan to Expand Drone Attacks," *Guardian*, March 19, 2009.

14. Loyd, "US Drone Strikes in Pakistan Tribal Areas."

15. "Pak May Say NO to Drone Hits in NWA," *Nation*, November 25, 2010.

16. Shane, "CIA to Expand Use of Drones in Pakistan."

17. "Drone War Can Trigger New Arms Race," *Arab News*, January 21, 2013, http://www.arabnews.com/drone-war-can-trigger-new-arms-race.

18. Josh Gerstein, "Dennis Blair Rips Obama White House," Politico.com, July 29, 2011, http://www.politico.com/news/stories/0711/60199.html.

19. Dennis Blair, "Drones Alone Are Not the Answer," *New York Times*, August 14, 2011.

20. Ken Dilanian, "US Put New Restrictions on CIA Drone Strikes in Pakistan," *Los Angeles Times*, November 7, 2011.

21. Nathaniel Fick, *From Strategy to Implementation: Strengthening US-Pakistani Relations* (Washington, DC: Center for a New American Security, July 7, 2009).

22. Masood, "Analysis: Time for a Rethink."

23. Entous, Gorman, and Barnes, "US Tightens Drone Rules."

24. "Pakistan Religious Party Condemns Drone Attacks," Agence France-Presse, January 27, 2012, http://www.google.com/hostednews/afp/article/ALeqM5iJQzDPFk7xuzE-YZfH0WOyThdFLA?docId=CNG.c1e6e8d2965a4dcf201447e6ac7d2984.b1.

25. Salman Siddiqui, "Drone Attacks Hit All Time High," *Express Tribune*, September 27, 2010.

26. David Rose, "How CIA Spies Deal Death from the Skies," *Daily Times*, April 17, 2011.

27. "Foreign Complicity in the Drone War," Al Jazeera, January 15, 2013, http://m.aljazeera.com/story/2013114103153393705.

28. Iftikhar Firdous, "Drone Attack Orphaned Whole Village," *Tribune Express*, December 22, 2011.

29. Ibid.

30. Nick Paton Walsh and Nasir Habib, "Pakistani Leaders Condemn Suspected US Drone Strike," CNN, March 18, 2011, http://articles.cnn.com/2011-03-18/world/pakistan.drone.strike_1_drone-strike-intelligence-officials-volatile-tribal-region.

31. Latif, "Pakistani Tribes Caught between Taliban and US Airstrikes."

32. Nadeem Sarwar, "US Drone War Brings Torment, Hope in Pakistan," EarthTimes.org, March 19, 2010.

33. Hasnain Kazim, "How the CIA Uses Pakistan as a Launch Pad for Drones," *Der Spiegel*, March 12, 2010.

34. Carol Grisanti and Mushtaq Yusufzai, "Pakistanis Outraged over Continued Drone Attacks," *World Blog*, NBC News, January 26, 2009, http://worldblog.nbcnews.com/_news/2009/01/26/4376849-pakistanis-outraged-over-continued-drone-attacks?lite.

35. Siddiqui, "Drone Attacks Hit All Time High."

36. Ibid.

37. Sudarsan Raghavan, "In Yemen Airstrikes Breed Anger, Sympathy for Al Qaeda," *Washington Post*, May 29, 2012.
38. Ibrahim Mothana, "How Drones Help Al Qaeda," *New York Times*, June 14, 2012.
39. Hakim Almasmari, "Two Suspected U.S. Drone Strikes Reported in Yemen," CNN, May 15, 2012, http://www.cnn.com/2012/05/15/world/meast/yemen-violence.
40. Sudarsan Raghavan, "When US Drone Strikes Kill Civilians, Yemen's Government Tries to Conceal It," *Washington Post*, December 24, 2012.
41. Sudarsan Raghavan, "In Yemen Airstrikes Breed Anger, Sympathy for Al Qaeda," *Washington Post*, May 29, 2012.
42. "Drone Strike Killed Americans," *RT*, October 17, 2011, http://rt.com/usa/news/drone-american-military-report-057/.
43. "Drones Kill Civilians," *Los Angeles Times*, May 29, 2010.
44. David Cloud, "Anatomy of an Afghan War Tragedy," *Los Angeles Times*, April 10, 2011.
45. Khaled Abdullah, "Drones Spur Yemenis' Distrust of Government and US," Reuters, October 27, 2010, http://www.reuters.com/article/2010/10/27/us-yemen-usa-qaeda-idUSTRE69Q36520101027.
46. United Press International, "US Drone Initiated Strike on Kurds," *Personal Liberty Digest*, January 8, 2012, http://personalliberty.com/2012/01/08/u-s-drone-initiated-strike-on-kurds/.
47. "Turkish Planes Kill 35 in Air Strike," Sky News, December 29, 2011, http://news.sky.com/story/912231/turkish-warplanes-kill-35-in-air-strike.
48. Oliver Farry, "Turkey Admits Mistake in Deadly Airstrike," France24, December 29, 2011, http://www.france24.com/en/20111229-turkey-admits-strike-over-kurdish-area-was-mistaken-smugglers-warplanes.
49. Hakim Almasmari, "Suspected US Drone Strike Kills Civilians in Yemen, Officials Say," CNN, September 3, 2012, http://www.cnn.com/2012/09/03/world/meast/yemen-drone-strike.
50. Mayer, "Predator War."
51. Orla Guerin, "Pakistani Civilian Victims Vent Anger over US Drones," BBC, November 3, 2011, http://www.bbc.co.uk/news/15553761.
52. Azmat Khan, "How the Drone War Plays Out in Pakistan's Tribal Areas," *Frontline*, PBS, December 12, 2011.
53. Ibid.
54. Justin Elliott, "I Wonder, Why I Was Victimized?" *Salon*, October 14, 2010, http://www.salon.com/2010/10/14/pakistan_civilian_deaths/.
55. Spencer Ackerman, "Rare Photographs Show Ground Zero of the Drone War," *Danger Room* (blog), *Wired*, December 12, 2011, http://www.wired.com/dangerroom/2011/12/photos-pakistan-drone-war/.
56. Saed Shan and Peter Beaumont, "US Drone Strikes in Pakistan Claiming Many Civilian Victims Says Campaigner," *Guardian*, July 17, 2011.
57. "Drones and Democracy," *Islam Times*, May 10, 2010.
58. Amna Nawaz, "For Many Pakistanis, 'USA' Means 'Drones,'" *World News* (blog), NBC News, June 26, 2012, http://worldnews.nbcnews.com/_news/2012/06/26/12403677-for-many-pakistanis-usa-means-drones.
59. Stanford Law School, "Victim's Stories," *Living under Drones*, 2012, http://livingunderdrones.org/victim-stories/.
60. Taj, "Drone Attacks: A Survey."
61. Gareth Porter, "CIA Slipping Its Leash with Drone Strikes," *Asia Times*, October 20, 2010.
62. Sarwar, "US Drone War Brings Torment."
63. Saeed Shah, "Pakistanis Protest Civilian Deaths in Drone Attacks," *Miami Herald*, December 10, 2010.
64. Smith, "Pakistan's Deadly Robots in the Sky."

65. Saed Hashmi, "Pakistani Tribesmen Ask for Sedatives Amid Stepped Up US Drone Strikes," *Xinhua*, September 27, 2010, http://news.xinhuanet.com/english2010/indepth/2010-09/27/c_13531919.htm.
66. Alex Rodriguez, "US Drone Attacks in Pakistan Get Mixed Response," *Los Angeles Times*, May 2, 2010.
67. Nawaz, "For Many Pakistanis, 'USA' Means 'Drones.'"
68. Jeb Boone, "Drone War: Strikes Fuel Anger in Yemen," *GlobalPost*, October 10, 2011, http://www.globalpost.com/dispatch/news/regions/middle-east/111007/drone-wars-yemen-unrest-protests-aqap-al-qaeda.

11. THE FUTURE OF KILLER DRONES

1. Charlie Savage, "Obama Adviser Discusses Using Military on Terrorists," *New York Times*, September 16, 2011.
2. Ahmed Rashid, "The US Pakistani Relationship in the Year Ahead," *Combating Terrorism Center Sentinel*, January 18, 2012.
3. Aamir Latif, "Obama's Hidden War: US Intensifies Drone Attacks in Pakistan," *GlobalPost*, May 23, 2011, http://www.globalpost.com/dispatch/news/regions/asia-pacific/pakistan/110520/al-qaeda-osama-bin-laden-north-waziristan.
4. "Charting the Data for US Air Strikes in Yemen."
5. Ibid.
6. Charlie Savage, "Awlaki Strike Shows US Shift to Drones in Terror Fight," *New York Times*, October 1, 2011.
7. Jim Michaels and Tom Brook, "Precision Strikes Are the New Weapon of Choice," *USA Today*, October 1, 2011.
8. Spencer Ackerman, "Obama's New Defense Plan: Drones, Spec Ops and Cyber War," *Danger Room* (blog), *Wired*, January 5, 2012, http://www.wired.com/dangerroom/2012/01/pentagon-asia-strategy/.
9. "Joe Biden Says America Needs More Drones, Gary Johnson Says the War Stops Here," *Washington Times*, October 12, 2012.
10. Julian Barnes, "More Drones, Fewer Troops," *Wall Street Journal*, January 26, 2012.
11. Julian Barnes, "US Expands Drone Flights to Take Aim at East Africa," *Wall Street Journal*, September 21, 2011; Jim Lobe, "US Expanding Network of Drone Bases to Hit Somalia, Yemen," Inter Press Service, September 21, 2011.
12. Craig Whitlock and Greg Miller, "US Assembling Secret Drone Bases in Africa, Arabian Peninsula Officials Say," *Washington Post*, September 20, 2011.
13. Greg Miller, "CIA Shifts Focus to Killing Targets," *Washington Post*, September 1, 2011; Noah Shachtman, "Is This the Secret US Drone Base in Saudi Arabia?" *Danger Room* (blog), *Wired*, February 7, 2013, http://www.wired.com/dangerroom/2013/02/secret-drone-base-2/.
14. Eric Schmitt, "US Troops at Drone Base in West Africa," *New York Times*, February 23, 2013.
15. Julian Barnes, "Floating Bases Enhance Capacity for Quick Strikes," *Wall Street Journal*, January 28, 2010.
16. Ibid.
17. Spencer Ackerman and Noah Shachtman, "Almost 1 in 3 US Warplanes Is a Robot," *Danger Room* (blog), *Wired*, January 9, 2012, http://www.wired.com/dangerroom/2012/01/drone-report/.
18. Greg Miller, "CIA to Expand Drone Fleet Officials Say," *Washington Post*, October 19, 2012.
19. "Flight of the Drones."
20. Rachel Martin, "Drone Pilots: The Future of Aerial Warfare," *Morning Edition*, National Public Radio, November 29, 2011, http://www.npr.org/2011/11/29/142858358/drone-pilots-the-future-of-aerial-warfare.
21. John Robb, "The Future of Drone Warfare," *Global Guerillas* (blog), December 21, 2011, http://globalguerrillas.typepad.com/globalguerrillas/2011/12/drone-bonjwas.html.

22. James Dao, "Drone Pilots Are Found to Get Stress Disorders Much as Those in Combat Do," *New York Times*, February 23, 2012.

23. "Drones and the Man."

24. Barnes, "More Drones, Fewer Troops."

25. "American Spy Drones Swarm in Droves over Afghanistan," *RT*, October 20, 2011, http://rt.com/usa/news/afghanistan-us-drones-nato-243/.

26. Ackerman and Shachtman, "Almost 1 in 3 US Warplanes."

27. Scott Shane, "Coming Soon: The Drone Arms Race," *New York Times*, October 8, 2011.

28. Nick Hopkins, "Afghan Civilians Killed by RAF Drone," *Guardian*, July 5, 2011.

29. Ibid.

30. "Taranis: The 143 Million Jet That Can Hit Targets on Another Continent," *Daily Mail*, July 13, 2010.

31. "United Arab Emirates Set to buy Predator Drones," *Los Angeles Times*, February 22, 2013.

32. "Pentagon Plans to Spend $37 Billion on Drones," Allgov.com, January 17, 2012, http://www.allgov.com/news/where-is-the-money-going/pentagon-plans-to-spend-37-billion-on-drones?news=843895.

33. Ibid.

34. "Armed UAV Operations 10 Years On," *Stratfor*, January 12, 2012, http://www.stratfor.com/weekly/armed-uav-operations-10-years.

35. Anwar Iqbal, "US Plans 75pc Increase in Drone Operations," *Dawn*, February 3, 2010, http://archives.dawn.com/archives/44399.

36. Ibid.

37. Robert Johnson, "The Navy's Modified X-47B Unmanned Drone Makes China's 'Carrier Killing' Missiles Obsolete," *Business Insider*, November 27, 2012, http://www.businessinsider.com/the-navys-modified-x-47b-unmanned-drone-makes-chinas-carrier-killing-missiles-obsolete-2012-11.

38. "Soon a Drone That Can Stay in the Air for 4 Days," *Times of India*, November 1, 2011.

39. David Hambling, "Air Force Completes Killer Micro-Drone Project," *Danger Room* (blog), *Wired*, January 5, 2010, http://www.wired.com/dangerroom/2010/01/killer-micro-drone/.

40. "US Army Orders First Suicide Drones," *Innovation News Daily*, September 6, 2011.

41. "Droning On," *Ottawa Citizen*, January 12, 2012.

42. Nick Turse, "Investigation Finds US Military Drones Have Flown Close to 3 Million Hours," *AlterNet*, November 1, 2011, http://www.alternet.org/story/152925/investigation_finds_u.s._military_drones_have_flown_close_to_3_million_hours.

43. "Faster Meaner Apaches Allow Pilots to Control Drones in Battle," Military.com, November 2, 2011.

44. "Predator B Drones Now Used to Monitor US Mexico Border," KSBY.com, September 1, 2010, http://www.ksby.com/news/predator-b-drones-now-used-to-monitor-u-s-mexico-border/#.

45. Associated Press, "FBI: Mass. Man Had 'Mission' against Pentagon," CBS News, November 4, 2011, http://www.cbsnews.com/8301-201_162-57319111/fbi-mass-man-had-mission-against-pentagon/.

46. Scott Wilson, "In Gaza, Life Shaped by Drones," *Washington Post*, December 3, 2011.

47. W. J. Hennigan, "The Air Force Buys an Avenger, Its Biggest and Fastest Armed Drone," *Los Angeles Times*, December 31, 2011.

48. Roggio and Mayer, "Charting the Data for US Airstrikes in Pakistan."

49. Willie Jones, "US Air Force Drone Touches Down amid Questions," *Tech Talk* (blog), *IEEE Spectrum*, December 6, 2010, http://spectrum.ieee.org/tech-talk/aerospace/military/us-air-force-drone-touches-down-amid-questions.

50. Adam Entous, "NATO to Acquire Fleet of Unmanned Drones," *Wall Street Journal*, February 4, 2012.

51. "NATO to Build Drone Base in Sicily," *Malta Star*, February 4, 2012.

52. Scott Wilson and Jon Cohen, "Poll Finds Broad Support for Obama's Counterterrorism Policies," *Washington Post*, February 7, 2012.

53. Tony Cappacio, "Pentagon to Increase Stock in High Altitude Drones," *Business Week*, February 5, 2010.

54. Jen Dimascio, "Pentagon Preps Attack of Drones," *Politico*, February 4, 2010, http://www .politico.com/news/stories/0210/32492.html.

55. "Drones in Houston Help Fight Iraq, Afghanistan Wars," *Houston Chronicle*, June 28, 2010.

56. It could be argued that Israel is already flying its killer drones abroad when it attacks Hamas militants in the Gaza Strip.

57. David Cortright, "The Scary Prospect of Global Drone Warfare," CNN, October 19, 2011, http://www.cnn.com/2011/10/19/opinion/cortright-drones.

58. W. J. Hennigan. "Civilian Use of Tiny Drones May Soon Fly in the US," *Los Angeles Times*, November 27, 2011.

APPENDIX: DRONE SPECIFICATIONS

1. U.S. Air Force, "MQ-1B Predator," fact sheet, January 5, 2012, http://www.af.mil/information /factsheets/factsheet.asp?fsID=122.

2. U.S. Air Force, "MQ-9 Reaper," fact sheet, January 5, 2012, http://www.af.mil/information /factsheets/factsheet.asp?fsID=6405.

SELECTED BIBLIOGRAPHY

Abbas, Hassan. *Pakistan's Drift into Extremism*. London: M. E. Sharpe, 2009.

Aid, Matthew. *Intel Wars: The Secret History of the Fight against Terror*. New York: Bloomsbury, 2012.

Ali, Imtiaz. "Pakistan Fury as CIA Airstrike on Village Kills 18." *Telegraph*, January 15, 2006.

Ali, Yousuf. "Most Bajaur Victims Were under 20." *News*, November 5, 2006.

Anderson, Terry. *Bush's Wars*. Oxford: Oxford University Press, 2011.

Bamford, James. *The Shadow Factory: The Ultra-Secret NSA from 9/11 to the Eavesdropping on America*. New York: Doubleday, 2009.

Bergen, Peter. *Holy War Inc*. New York: Touchstone, 2001.

———. *The Longest War: The Enduring Conflict between America and Al-Qaeda*. New York: Free Press, 2011.

Bergen, Peter, and Katherine Tiedemann. *The Drone War*. Washington, DC: New America Foundation, June 3, 2009.

———. *The Year of the Drone: An Analysis of Drone Strikes in Pakistan, 2004–2011*. Washington, DC: New America Foundation, August 2, 2011. http://counterterrorism.newamerica.net/drones.

Boone, Jon, and Jason Burke. "Top Al Qaeda Leader Killed in Afghanistan." *Guardian*, January 31, 2008.

Brook, Tom. "US Shifts Spy Planes to Afghan War." *USA Today*, August 23, 2009.

Burke, Jason. "Al Qaeda Chief Dies in Missile Strike." *Guardian*, June 1, 2008.

Clarke, Richard. *Against All Enemies: Inside America's War on Terror*. New York: Free Press, 2004.

Cloud, David. "CIA Drones Have Broader List of Targets." *Los Angeles Times*. May 5, 2010.

Coll, Steve. *Ghost Wars: The Secret History of the CIA, Afghanistan, and Bin Laden, from the Soviet Invasion to September 10, 2001*. New York: Penguin, 2005.

Darling, Dan. "Al Qaeda's Mad Scientist." *Weekly Standard*, January 19, 2006.

Dilanian, Ken. "CIA Drones May Be Avoiding Pakistani Civilians." *Los Angeles Times*, February 22, 2011.

Ensor, David. "Drone May Have Spotted Bin Laden in 2000." CNN, March 17, 2004. http://articles.cnn.com/2004-03-17/world/predator.video_1_bin -senior-al-predator-drone-aircraft.

———. "Sources: Key Al Qaeda Operative Killed." CNN, May 14, 2005. http://www.cnn.com/2005/WORLD/asiapcf/05/13/alqaeda.killing/index.html.

Entous, Adam, Siobhan Gorman, and Julian Barnes. "US Tightens Drone Rules." *Wall Street Journal*, November 4, 2011.

Fair, Christine. "Drones over Pakistan—Menace or Best Viable Option?" *The Blog* (blog), *Huffington Post*, August 2, 2010. http://www.huffingtonpost.com/c -christine-fair/drones-over-pakistan——m_b_666721.html.

Gillmor, Dan. "Suicide Blast Kills 42 Pakistani Soldiers." *Guardian*, November 8, 2006.

Gul, Imtiaz. *The Most Dangerous Place: Pakistan's Lawless Frontier*. New York: Viking, 2010.

Kelly, Mary Louise. "Bin Laden Son Reported Killed in Pakistan." National Public Radio, July 22, 2009. http://www.npr.org/templates/story/story.php?storyId =106903109.

Khan, Anwarullah. "12 Killed in Drone Attack on Damadola." *Dawn*, May 14, 2008.

Khan, S. H. "US Strike Kills Eight Taliban in Pakistan, Officials." Agence France-Presse, August 27, 2009.

Klaidman, Daniel. *Kill or Capture: The War on Terror and the Soul of the Obama Presidency*. Boston: Houghton Mifflin Harcourt, 2012.

Lamb, Christina. "US Carried Out Madrasah Bombing." *Sunday Times*, November 26, 2006.

Masood, Salman. "Video of Flogging Rattles Pakistan." *New York Times*, April 4, 2009.

Mayer, Jane. "The Predator War: What Are the Risks of the C.I.A.'s Covert Drone Program?" *New Yorker*, October 26, 2009.

Miller, Greg. "Increased Drone Strikes in Pakistan Killing Few High Value Militants." *Washington Post*, February 21, 2011.

———. "War on Terror Loses Ground: Al Qaeda Is Regrouping in Pakistan, an Ally the U.S. Must Work with but Doesn't Trust." *Los Angeles Times*, July 27, 2008.

Mockenhaupt, Brian. "We've Seen the Future and Its Unmanned." *Esquire*, October 14, 2009.

Moreau, Ron, et al. "Day of the Vampire." *Newsweek*, June 17, 2002.

Perlez, Jane, and Pir Zubair Shah. "US Attack on Taliban Kills 23 in Pakistan." *New York Times*, September 9, 2009.

Plaw, Avery. *Targeting Terrorists: A License to Kill?* Burlington, VT: Ashgate, 2008.

Rashid, Ahmed. *Descent into Chaos: The US and the Failure of Nation Building in Pakistan, Afghanistan, and Central Asia*. New York: Viking, 2008.

Reeves, Philip. "Did Baitullah Mehsud Kill Benazir Bhutto?" *All Things Considered*, National Public Radio, January 16, 2008. http://www.npr.org /templates/story/story.php?storyId=18159635.

Rennie, David. "US Tried to Kill Warlord with Drone Missile." *Telegraph*, May 10, 2002.

Riedel, Bruce. *Deadly Embrace: Pakistan, America, and the Future of Global Jihad.* Washington, DC: Brookings Institution, 2012.

Roggio, Bill. "Cross-Border Strike Targets One of the Taliban's 157 Training Camps in Pakistan's Northwest." *Long War Journal*, August 13, 2008. http://www.longwarjournal.org/archives/2008/08/crossborder_strike_t.php.

———. "Pakistan Implicates Baitullah Mehsud in Bhutto Assassination." *Long War Journal*, December 28, 2007. http://www.longwarjournal.org/archives/2007 /12/pakistan_implicates.php.

———. "Pakistani Leader Killed in March 2008 Predator Strike." *Long War Journal*, May 19, 2008. http://www.longwarjournal.org/archives/2009/05/pakistani _al_qaeda_l.php.

———. "The Pro-Osama Meeting in Bajaur." *Long War Journal*, October 29, 2006. http://www.longwarjournal.org/archives/2006/10/the_proosama_meeting .php.

———. "Report: US Airstrike Kills 4 in North Waziristan." *Long War Journal*, September 4, 2008. http://www.longwarjournal.org/archives/2008/09/report _us_airstrike.php

———. "Scores of Taliban Killed in Second US Strike in South Waziristan." *Long War Journal*, June 23, 2009. http://www.longwarjournal.org/archives/2009 /06/seventeen_taliban_ki.php.

Roggio, Bill, and Alexander Mayer. "Charting the Data for US Airstrikes in Pakistan, 2004–2011." *Long War Journal*. http://www.longwarjournal.org /pakistan-strikes.php. Accessed December 2011.

Rohde, David, and Kristen Mulvihill. *A Rope and a Prayer: A Kidnapping from Two Sides.* New York: Viking, 2010.

Ross, Brian. "U.S. Strike Killed Al Qaeda Bomb Maker." ABC, January 18, 2006. http://abcnews.go.com/WNT/Investigation/story?id=1517986.

Rubin, Elizabeth. "In the Land of the Taliban." *New York Times*, October 22, 2006.

Sanger, David E. *The Inheritance: The World Obama Confronts and the Challenges to American Power.* New York: Harmony Books, 2009.

Schmidt, John. *The Unraveling: Pakistan in the Age of Jihad.* New York: Farrar, Strauss and Giroux, 2011.

Schmitt, Eric, and Thom Shanker. *Counterstrike: The Untold Story of America's Secret Campaign against Al Qaeda.* New York: Time Books, 2011.

Shachtman, Noah. "No-Name Terrorists Now CIA Drone Targets." *Danger Room* (blog), *Wired*, May 6, 2010. http://www.wired.com/dangerroom/2010/05 /no-name-terrorists-now-cia-drone-targets/.

Shane, Scott. "CIA to Expand Use of Drones in Pakistan." *New York Times,* December 3, 2008.

Shannon, Elaine. "Can Bin Laden Be Caught?" *Time,* January 22, 2006.

Siddiqui, Tayyab. "Pakistan's Drone Dilema." *Dawn,* July 18, 2010.

Singer, Peter. *Wired for War: The Robotics Revolution and Conflict in the Twenty-First Century.* New York: Penguin Books, 2009.

Tavernise, Sabrina. "Deaths at Hands of Militants Rise in Pakistan." *New York Times,* January 14, 2009.

Tenet, George. *At the Center of the Storm: My Years at the CIA.* New York: HarperCollins, 2007.

Thirtle, Michael R., Robert Johnson, and John Birkler. *The Predator ACTD: A Case Study for Transition Planning to the Formal Acquisition Process.* Santa Monica, CA: Rand, 1997.

Vanden Brook, Tom. "Faster, Deadlier Pilotless Plane Bound for Afghanistan." *USA Today,* August 27, 2007.

Walsh, Declan. "Mysterious 'Chip' Is CIA's Latest Weapon against Al-Qaida Targets Hiding in Pakistan's Tribal Belt." *Guardian,* May 31, 2009.

———. "US Bomb Kills Eleven Pakistani Troops." *Guardian,* June 12, 2008.

Weaver, Mary Anne. *Pakistan: Deep Inside the World's Most Frightening State.* New York: Farrar, Straus and Giroux, 2010.

Williams, Brian Glyn. *Afghanistan Declassified: A Guide to America's Longest War.* Philadelphia: University of Pennsylvania Press, 2012.

———. "Cheney Attack Reveals Taliban Suicide Bomber Targeting Patterns." *Terrorism Monitor* 5, no. 4 (March 1, 2007): 4–7.

———. "The CIA's Covert Predator Drone War in Pakistan, 2004–2010: The History of an Assassination Campaign." *Studies in Terrorism and Conflict* 33 (2010): 871–92.

———. *The Last Warlord: The Life and Legend of Dostum, the Afghan Warrior who Led US Special Forces to Topple the Taliban Regime.* Chicago: Chicago Review Press, 2013.

———. "New Light on the Accuracy of the CIA's Predator Drone Campaign in Pakistan." *Terrorism Monitor* 8, no. 41 (November 11, 2010).

———. "Report from the Field: General Dostum and the Mazar i Sharif Campaign: New Light on the Role of Northern Alliance Warlords in Operation Enduring Freedom." *Small Wars and Insurgencies* 21, no. 4 (December 2010): 12–25.

Woodward, Bob. *Obama's Wars.* New York: Simon & Schuster, 2010.

Yenne, Bill. *Birds of Prey: Predators, Reapers and America's Newest UAVs in Combat.* North Branch, MN: Specialty Books, 2010.

Yusuf, Huma. "Fallout of the Davis Case." *Dawn,* February 21, 2011.

Zaidi, Hassan. "Army Chief Wanted More Drone Support." *Dawn,* May 20, 2011.

INDEX

ABOUT THE AUTHOR

Brian Glyn Williams is professor of Islamic history at the University of Massachusetts–Dartmouth and previously taught at the School of Oriental and African Studies at the University of London. He has a PhD in Central Asian history and master's degrees in Central Eurasian studies and in Russian history. He has extensive travel experience in Afghanistan, where he lived with Northern Alliance warlords, worked for the U.S. military, and tracked suicide bombers for the CIA Counterterrorism Center. Most recently he published a civilian version of a field manual he wrote for the U.S. Army's Joint Information Operations Warfare Command titled *Afghanistan Declassified: A Guide to America's Longest War*. He has also carried out fieldwork in Pakistan on the CIA's drone assassination campaign. His forthcoming book is *The Last Warlord: The Life and Legend of Dostum, the Afghan Warrior Who Led US Special Forces to Topple the Taliban Regime*. For articles, videos, and photographs from his fieldwork, see his website: brianglynwilliams.com.